基于Kotlin的
Spring Boot微服务实战

袁康 / 著

内 容 简 介

本书介绍了 Kotlin 在 Spring Boot 微服务开发中的实践，并使用 Kotlin 作为开发语言，介绍了函数式编程思想、Kotlin 的语法、Kotlin 在常用中间件中的应用，以及其在微服务注册中心、微服务配置中心、微服务网关、Spring Cloud Alibaba、服务监控和服务链路监控方面的应用。本书给出了详细的实例代码和一个完整的博客示例，可以帮助读者使用 Kotlin 开发基于 Spring Boot 微服务的程序。

阅读本书需要具有一定的编程基础，但入门门槛不高。因此，本书适合大学生、工程师等对使用 Kotlin 开发服务端程序感兴趣的读者阅读。

未经许可，不得以任何方式复制或抄袭本书之部分或全部内容。
版权所有，侵权必究。

图书在版编目（CIP）数据

基于 Kotlin 的 Spring Boot 微服务实战/袁康著. —北京：电子工业出版社，2020.11
ISBN 978-7-121-39715-8

Ⅰ.①基… Ⅱ.①袁… Ⅲ.①JAVA 语言－程序设计 Ⅳ.①TP312.8

中国版本图书馆 CIP 数据核字（2020）第 189280 号

责任编辑：付　睿
印　　刷：三河市君旺印务有限公司
装　　订：三河市君旺印务有限公司
出版发行：电子工业出版社
　　　　　北京市海淀区万寿路 173 信箱　　　邮编：100036
开　　本：787×980　1/16　　印张：24　　字数：509 千字
版　　次：2020 年 11 月第 1 版
印　　次：2020 年 11 月第 1 次印刷
定　　价：99.00 元

凡所购买电子工业出版社图书有缺损问题，请向购买书店调换。若书店售缺，请与本社发行部联系，联系及邮购电话：（010）88254888，88258888。
质量投诉请发邮件至 zlts@phei.com.cn，盗版侵权举报请发邮件至 dbqq@phei.com.cn。
本书咨询联系方式：（010）51260888-819，faq@phei.com.cn。

推荐序

编程语言始终是程序员群体中最容易引起争论的话题。但要说起近年来最受程序员欢迎的语言，Kotlin 一定可以占据一席之地。在 TOIBE 和 RedMonk 编程语言排名中，Kotlin 已经跻身 JVM 语言前列，和老牌的 Scala 和 Groovy 形成三足鼎立之势。在权威的 ThoughtWorks 技术雷达中，Kotlin 从 2017 年 3 月首次入选，到 2018 年 5 月变成推荐在实际生产环境中采纳，只用了一年多的时间。Kotlin 语言究竟有什么魔力能够迅速获得广大程序员的青睐呢？

首先，Kotlin 语言具备优秀、简洁的特性。它从一开始就保留了 Java 语言的静态类型，加入了诸如空安全、扩展函数、数据类、Lambda 表达式等现代编程语言的功能。随着 Kotlin 版本的不断升级，跨平台原生、协程等优秀的语言特性不断地扩展着该语言的功能和适用场景。而且，Kotlin 语言因其与 JVM 语言的良好互操作性和简洁的语法，可以在采用 Java 语言的依赖项目上快速地被无缝引入，大大降低了新编程语言的学习成本。

其次，Kotlin 的发展离不开大厂的鼎力支持和成功案例的支撑。在 2017 年的 Google I/O 大会上，Kotlin 成为 Android 开发的首选语言。而经过两年的发展，Google 提供的 Android Jetpack 扩展都针对 Kotlin 语言进行了改写和升级，编译工具 Gradle 也增加了对 Kotlin 编写脚本的支持。服务端开发的主流框架 Spring 几乎在第一时间就支持了 Kotlin 语言，且在不断地丰富 Spring 的 Kotlin 扩展。Kotlin 语言已经在移动端、前端、服务端、科学计算等领域得到了广泛应用。

最后，开发者有口皆碑的语言一定会在开发者社区中引起共鸣。JetBrains 作为 Kotlin 语言的缔造者，在语言特性的引入上始终都非常尊重开发者社区的反馈，在 API 的设计上也十分慎重。国内的 Kotlin 社区几乎和 Kotlin 语言同时诞生，相关的文档、资料也是第一时间就能得到翻译和推广。更难能可贵的是，国内很多 Kotlin 开发者将自己的经验进行总结并分享出来。本书就是一本国内开发者在探索 Kotlin 实践中的的优秀总结。

不同于市面上专注于语言特性或者 Android 开发的 Kotlin 图书，本书将 Kotlin 语言在服务端中的应用作为重点。基于 Java 语言的 Spring 框架是服务端开发领域的主流选择，而且国内开发者也在 Spring 框架的基础上增加了 Dubbo 这样的创新。本书作者将 Kotlin 和 Spring 结合的实战经验对广大国内服务端开发者来说是不可多得的参考资料。初次接触 Kotlin 语言的读者也不必担心，本书还包括了对 Kotlin 语言的基本特性和开发工具的介绍，可以帮助读者快速进入状态。

感谢作者为国内广大开发者带来了一本新鲜的 Kotlin 实战手册！希望读者能够感受到 Kotlin 语言的独特魅力！

《Kotlin 实战》译者，Thoughtworks 资深咨询师

覃宇

读者服务

微信扫码回复：39715

- 获取本书配套资源[1]
- 获取作者提供的各种共享文档、线上直播、技术分享等免费资源
- 加入读者交流群，与更多读者互动
- 获取博文视点学院在线课程、电子书 20 元代金券

---

1 请访问 http://www.broadview.com.cn/39715 下载本书提供的附加参考资料，如正文中提及参见"链接1""链接2"等时，可在下载的"参考资料.pdf"文件中查阅。

序言

近年来，Java 增加了函数式编程的特性，如类型推断、Lambda 表达式、Stream 流等，后端开发逐渐采用了一些函数式编程语言，如 Scala、Kotlin。函数式编程以其简捷性、不变性、空指针处理友好等特点深受后端开发人员的青睐。

笔者之前使用 Scala 做过后端开发，感慨 Scala 的门槛较高，入门困难，而且生态资源相对匮乏，开发过程比较痛苦。后来，使用 Java 进行后端开发，依托 Spring Boot 强大的生态，可以方便地使用消息队列、数据库、缓存、大数据相关中间件。但是 Java 比较笨重，冗余的代码、空指针异常、线程安全等问题常常困扰着笔者。随着 Kotlin 在移动端开发的普及，它也逐步走入后端开发者的视野。Kotlin 是 JVM 体系的语言，和 Java 有着良好的互操作性，上手较容易，且可以使用 Java 强大的生态，其还具有函数式编程的优点。另外，Spring Initializr 提供了对 Java、Kotlin 语言的支持。

Kotlin 是 JetBrains 公司开发的，目前流行的 IntelliJ IDEA 软件也是该公司开发的。IDEA 对 Kotlin 支持较好，可以将 Java 代码转换为 Kotlin 代码。IDEA 还支持 Java、Kotlin 混合编程，历史代码使用 Java 编写，新的代码可以尝试使用 Kotlin 编写。

基于以上考虑，笔者开始研究使用 Kotlin、Spring Boot 做后端开发，取得了不错的效果。市面上介绍使用 Kotlin 进行后端开发的图书比较少，笔者在大量实践的基础上，萌生了写一本书的想法，希望和更多的 Java 开发人员分享 Kotlin 在后端开发中的实践经验。

本书共 10 章，第 1 章介绍如何搭建 Kotlin 的开发环境，第 2 章介绍函数式编程，第 3 章简单介绍 Kotlin 的语法，第 4 章介绍 Kotlin 在常用中间件中的应用，第 5 章介绍 Kotlin 如何应用于微服务注册中心，第 6 章介绍 Kotlin 如何应用于微服务配置中心，第 7 章介绍 Kotlin 如何应用于微服务网关，第 8 章介绍 Kotlin 如何应用于 Spring Cloud Alibaba，第 9 章介绍 Kotlin 集成服务监控和服务链路监控的相关知识，第 10 章介绍如何用 Kotlin 编写博客应用。本书提供了大量的实例，相关源码可以从 GitHub 下载运行。

袁康

2020 年 5 月 26 日于上海

目录

第 1 章　搭建 Kotlin 开发环境 1
  1.1　Kotlin 简介 1
  1.2　在 Windows 环境中搭建 Kotlin 开发环境 5
  1.3　在 Ubuntu 环境中搭建 Kotlin 开发环境 9
  1.4　在 macOS 环境中搭建 Kotlin 开发环境 12
  1.5　第一个 Kotlin 程序 13
  1.6　小结 14

第 2 章　函数式编程介绍 15
  2.1　初识函数式编程 15
  2.2　函数式编程的特点 17
  2.3　Scala、Kotlin、Java 的对比 20
  2.4　小结 21

第 3 章　Kotlin 的语法 22
  3.1　基础语法 22
    3.1.1　基本数据类型 22
    3.1.2　包名和引用 27
    3.1.3　流程控制 28
    3.1.4　返回和跳转 29
  3.2　类 30
    3.2.1　类、属性、接口 30
    3.2.2　特殊类 34

  3.2.3 泛型 36
  3.2.4 委托 38
 3.3 函数和 Lambda 表达式 40
  3.3.1 函数 40
  3.3.2 Lambda 表达式 42
 3.4 集合 44
  3.4.1 集合概述 44
  3.4.2 集合操作 51
  3.4.3 List、Set、Map 相关操作 58
 3.5 协程 60
  3.5.1 协程基础 60
  3.5.2 协程进阶 64
 3.6 小结 67

第 4 章 Kotlin 在常用中间件中的应用 68

 4.1 Kotlin 集成 Spring Boot 68
  4.1.1 Spring Boot 介绍 68
  4.1.2 用 Kotlin 开发一个 Spring Boot 项目 69
 4.2 Kotlin 集成 Redis 73
  4.2.1 Redis 介绍 74
  4.2.2 使用 Kotlin 操作 Redis 75
 4.3 Kotlin 集成 JPA、QueryDSL 81
  4.3.1 JPA、QueryDSL 介绍 82
  4.3.2 使用 Kotlin 操作 JPA、QueryDSL 83
 4.4 Kotlin 集成 MongoDB 91
  4.4.1 MongoDB 介绍 91
  4.4.2 使用 Kotlin 操作 MongoDB 92
 4.5 Kotlin 集成 Spring Security 98
  4.5.1 Spring Security 介绍 98
  4.5.2 使用 Kotlin 操作 Spring Security 100

- 4.6 Kotlin 集成 RocketMQ 105
  - 4.6.1 RocketMQ 介绍 105
  - 4.6.2 使用 Kotlin 操作 RocketMQ 107
- 4.7 Kotlin 集成 Elasticsearch 112
  - 4.7.1 Elasticsearch 介绍 112
  - 4.7.2 使用 Kotlin 操作 Elasticsearch 113
- 4.8 Kotlin 集成 Swagger 119
  - 4.8.1 Swagger 介绍 119
  - 4.8.2 使用 Kotlin 操作 Swagger 121
- 4.9 小结 126

第 5 章 Kotlin 应用于微服务注册中心 127

- 5.1 Eureka 127
  - 5.1.1 Eureka 介绍 127
  - 5.1.2 Kotlin 集成 Eureka 服务注册 129
  - 5.1.3 一个 Eureka 服务提供方 131
  - 5.1.4 Kotlin 集成 OpenFeign 服务调用 135
  - 5.1.5 Kotlin 集成 Ribbon 服务调用 139
- 5.2 Consul 143
  - 5.2.1 Consul 介绍 144
  - 5.2.2 Kotlin 集成 Consul 服务注册 145
  - 5.2.3 Kotlin 集成 OpenFeign 和 Ribbon 服务调用 149
- 5.3 Zookeeper 153
  - 5.3.1 Zookeeper 介绍 153
  - 5.3.2 Kotlin 集成 Zookeeper 服务注册 154
  - 5.3.3 Kotlin 集成 OpenFeign 和 Ribbon 服务调用 158
- 5.4 Nacos 163
  - 5.4.1 Nacos 介绍 163
  - 5.4.2 Kotlin 集成 Nacos 服务注册 164
  - 5.4.3 Kotlin 集成 OpenFeign 和 Ribbon 服务调用 167
- 5.5 小结 171

第 6 章　Kotlin 应用于微服务配置中心 172

6.1　Spring Cloud Config 172
6.1.1　Spring Cloud Config 介绍 172
6.1.2　Kotlin 集成 Spring Cloud Config 173

6.2　Apollo 配置中心 181
6.2.1　Apollo 介绍 181
6.2.2　Kotlin 集成 Apollo 182

6.3　Nacos 配置中心 186
6.4　Consul 配置中心 192
6.5　小结 197

第 7 章　Kotlin 应用于微服务网关 198

7.1　Kotlin 集成 Zuul 198
7.1.1　Zuul 介绍 198
7.1.2　Kotlin 集成 Zuul 200

7.2　Kotlin 集成 Spring Cloud Gateway 211
7.2.1　Spring Cloud Gateway 介绍 211
7.2.2　Kotlin 集成 Spring Cloud Gateway 212

7.3　小结 222

第 8 章　Kotlin 应用于 Spring Cloud Alibaba 223

8.1　服务限流降级 224
8.1.1　Sentinel 介绍 225
8.1.2　Kotlin 集成 Sentinel 226

8.2　消息驱动 232
8.2.1　消息驱动介绍 233
8.2.2　Kotlin 集成 RocketMQ 实现消息驱动 234

8.3　阿里对象云存储 241
8.3.1　阿里对象云存储介绍 241
8.3.2　Kotlin 集成阿里对象云存储 242

8.4 分布式任务调度 248
    8.4.1 SchedulerX 介绍 248
    8.4.2 Kotlin 集成 SchedulerX 249

8.5 分布式事务 253
    8.5.1 分布式事务介绍 253
    8.5.2 Kotlin 集成 Seata 255

8.6 Spring Cloud Dubbo 270
    8.6.1 Dubbo 介绍 270
    8.6.2 Kotlin 集成 Spring Cloud Dubbo 271

8.7 小结 279

第 9 章 Kotlin 集成服务监控和服务链路监控 280

9.1 Prometheus、Grafana 介绍 280

9.2 Kotlin 集成 Prometheus、Grafana 282

9.3 Kotlin 集成 Zipkin 288

9.4 Kotlin 集成 SkyWalking 298

9.5 小结 308

第 10 章 基于 Kotlin 和 Spring Boot 搭建博客 309

10.1 初始化 Maven 工程 309

10.2 系统架构 314

10.3 定义实体 316

10.4 数据库设计 324

10.5 Repository 层的设计 325

10.6 Service 层的设计 346

10.7 Controller 层的设计 353

10.8 部署到腾讯云 370

10.9 小结 373

第 1 章 搭建 Kotlin 开发环境

本章主要介绍如何在 Windows、Linux、macOS 平台搭建 Kotlin 开发环境，包括安装 JDK、IDEA。本章还将简单介绍 Kotlin 语言的特性，Kotlin 是一门运行在 Java 平台的函数式编程语言。在本章最后将使用 IDEA 编写一个 Kotlin 示例程序。

1.1 Kotlin 简介

Kotlin 是一门运行在 Java 平台的函数式编程语言，由 JetBrains 公司开发。Kotlin 支持多个平台，包括移动端、服务端及浏览器端。Kotlin 历经了 1.1、1.2 版本，目前最新的版本是 1.3。

Kotlin 融合了面向对象和函数式编程，其简洁、安全、优雅，可以和 Java 完全交互，并可使用 Java 编写的第三方 Jar 包。下面我们以一个小例子解释 Kotlin 是什么。这个例子定义了一个 Animal 类，然后创建了一些 Animal 类的对象，要找出其中 age 最大的 Animal 对象并将其打印出来。在这个例子中，可以看到 Kotlin 的许多特性。代码如下：

```
1.  //1 数据类
2.  data class Animal(val name: String,
3.                    //2 可为空的类型(Int?)，声明变量的默认值
4.                    val age: Int? = null)
```

```
5.
6.    //3 顶层函数
7.    fun main(args: Array<String>) {
8.        //4 命名声明
9.        val animals = listOf(Animal("Dog"), Animal("Cat", 3))
10.
11.       //5 Lambda 表达式
12.       val oldest = animals.maxBy { it.age ?: 0 }
13.       //6 字符串模板
14.       println("The Oldest is $oldest")
15.
16.       //7 自动生成 toString 方法
17.       //The Oldest is Animal(name=Cat, age=3)
18.   }
```

这里声明了一个 Animal 对象，它带有两个属性 name 和 age。age 属性的默认值为 null。当创建一个 name=Dog 的 Animal 对象时，没有设置 age 属性，默认值是 null。之后，使用 maxBy 方法查找年龄最大的 Animal 对象，使用 it 作为默认的参数名。如果 age 是 null 的话，设置默认值为 0。由于没有指定 Dog 的年龄，因此使用默认值 0。最终，Cat 是年龄最大的 Animal 对象。

通过这个例子，我们对 Kotlin 有了一个初步的印象，下面介绍 Kotlin 的几个特性。

静态类型

Kotlin 是一种静态类型编程语言，编译器可以验证程序中的变量、表达式类型。而基于 JVM 的动态类型编程语言，如 Groovy，可以在运行时解析方法和字段引用。另外，Kotlin 不需要在代码中指定每一个变量的类型，在许多场景中，能够根据上下文自动推断变量类型。目前大多数公司仍然在使用 Java 7 或 Java 8，这些 Java 版本没有这个特性。下面是一个简单的例子：

```
Val x = 1
```

其中声明了一个变量，由于它以一个整数初始化，因此 Kotlin 会自动推断这个变量的类型为 Int。静态类型有如下好处。

- **性能**：由于不需要在运行时判断需要调用哪个方法，因此方法调用速度快。
- **可读性**：由于编译器校验了程序的正确性，因此在运行时发生崩溃的可能性较小。
- **可维护性**：由于能看到代码调用了什么类型的对象，所以可以更容易地理解代码。
- **工具支持**：有丰富的 IDE 支持编程。

Kotlin 支持可为空的类型（nullable type），可以在编译时检查可能的空指针异常，使得程序更加可靠。此外，Kotlin 还支持函数类型（functional type）。

支持函数式编程和面向对象编程

函数式编程的关键概念如下：

- **函数是一等公民**。可以把函数看作一个值，把函数存储在变量中作为一个参数传递或者返回。
- **不变性**。使用不可变的对象，一旦对象被创建，它的状态不可更改。
- **没有副作用**。给定相同的输入将会返回相同的结果。它不会修改其他对象的状态或者和外界进行交互

用函数编写代码简洁、优雅。将函数作为一个值可带来更强大的抽象力，避免代码冗余。函数式编程是线程安全的。如果使用不可修改的数据结构和纯函数，可以确保不会出现不安全的修改，也不需要设计复杂的同步方案。Kotlin 有丰富的特性支持函数式编程：

- 允许函数接收其他函数作为参数或者返回其他函数。
- 使用 Lambda 表达式，使用最小的模板分发代码块。
- 为创建不可变对象提供了精简的语法。
- 标准库为以函数式风格使用对象和集合提供了丰富的 API。

实用性

Kotlin 基于多年的大规模系统设计的行业经验，借鉴了许多软件开发者遇到的案例，因此具有极强的实用性。JetBrains 公司和社区的开发者已经使用 Kotlin 早期版本许多年，他们的各种反馈已经融合到发行版中。Kotlin 不强迫你使用任何特定的编程风格或范式。当你使用这门语言时，可以使用自己在进行 Java 开发时熟悉的风格。IntelliJ IDEA 提供了对 Kotlin 强大的支持，可自动将 Java 转换为 Kotlin，可以优化代码块，并提供了代码填充功能。

精简

代码越精简，人理解得越快，维护越方便。使用 Kotlin，可以简便地定义数据类，省略了大量的 getter、setter 和将构造器参数赋值给字段的逻辑。Kotlin 拥有丰富的标准库，这些库让你可以通过调用库函数来代替那些冗长的、重复的代码片段。Kotlin 对 Lambda 表达式的支持使得将少部分代码传递到库函数变得十分容易。这让你可以将所有通用的部分封装到函数库，代码中仅保留业务相关的部分。

安全

使用 Kotlin 可以通过花费较小的代价即达到一个比 Java 更高级别的安全水平。Kotlin 努力从程序中移除空指针异常，其类型系统跟踪可能为空的值，并且禁止运行时导致空指针异常的操作。它需要的额外成本是最小的：只需要一个单独的字符，在末尾加一个问号，即标记为一个可能为空的类型。

```
1.  val s: String? =   null      // 1  可能为空
2.  val s2: String =   ""        // 2  一定不为空
```

另外，Kotlin 提供了许多便捷的方式来处理可能为空的数据，这在避免应用程序崩溃方面提供了极大的帮助。

Kotlin 有助于避免类型强转异常。使用 Java 编程，开发者经常会遗漏类型检查，而 Kotlin 将类型检查和转换合并为一个单独的操作，一旦检查了类型，就可以应用该类型的成员而无须额外进行类型转换。

```
1.  if(value is String)          // 1 类型检查
2.      println(value.toUpperCase())   // 2 调用该类型的方法
```

互操作性

Kotlin 可以使用已有的 Java 库，调用 Java 方法，扩展 Java 类，实现 Java 接口。同时，也可以在 Java 代码中调用 Kotlin 代码。Kotlin 的类和方法能够像常规的 Java 类和方法那样被调用，而且我们还可以在项目的任何地方混合使用 Java 和 Kotlin。

Kotlin 尽可能地使用现有的 Java 库。例如，Kotlin 并没有自己的集合库，它依赖 Java 标准库的类，通过额外的函数来扩展它们，这样就可以更加方便地使用 Kotlin 了。

Kotlin 工具还为多语言项目提供了全面的支持。它能够编译任意一个混合 Java 和 Kotlin 的源文件，不论它们之间是如何相互依赖的。这个 IDE 特性对其他语言也是有效的。它将允许你做以下事情：

- 在 Java 和 Kotlin 源文件中自由切换。
- 调试混合语言项目并在用不同语言编写的代码中进行单步跟踪。
- 使用 Kotlin 重构和正确地升级你的 Java 函数。

1.2　在 Windows 环境中搭建 Kotlin 开发环境

搭建 Kotlin 开发环境需要安装 Java、IntelliJ IDEA 工具。本书使用 Java 1.8 和 IntelliJ IDEA 2019.2.4 社区版本。IntelliJ IDEA 自带 Kotlin 插件（Kotlin 的版本是 1.3.60）。

要在Oracle官网 [1]下载Java SE Development Kit 8u231，只需单击下载jdk-8u231-windows-x64.exe文件，如图 1.1 所示。

Java SE Development Kit 8u231

You must accept the Oracle Technology Network License Agreement for Oracle Java SE to download this software.

Thank you for accepting the Oracle Technology Network License Agreement for Oracle Java SE; you may now download this software.

Product / File Description	File Size	Download
Linux ARM 32 Hard Float ABI	72.9 MB	jdk-8u231-linux-arm32-vfp-hflt.tar.gz
Linux ARM 64 Hard Float ABI	69.8 MB	jdk-8u231-linux-arm64-vfp-hflt.tar.gz
Linux x86	170.93 MB	jdk-8u231-linux-i586.rpm
Linux x86	185.75 MB	jdk-8u231-linux-i586.tar.gz
Linux x64	170.32 MB	jdk-8u231-linux-x64.rpm
Linux x64	185.16 MB	jdk-8u231-linux-x64.tar.gz
Mac OS X x64	253.4 MB	jdk-8u231-macosx-x64.dmg
Solaris SPARC 64-bit (SVR4 package)	132.98 MB	jdk-8u231-solaris-sparcv9.tar.Z
Solaris SPARC 64-bit	94.16 MB	jdk-8u231-solaris-sparcv9.tar.gz
Solaris x64 (SVR4 package)	133.73 MB	jdk-8u231-solaris-x64.tar.Z
Solaris x64	91.96 MB	jdk-8u231-solaris-x64.tar.gz
Windows x86	200.22 MB	jdk-8u231-windows-i586.exe
Windows x64	210.18 MB	jdk-8u231-windows-x64.exe

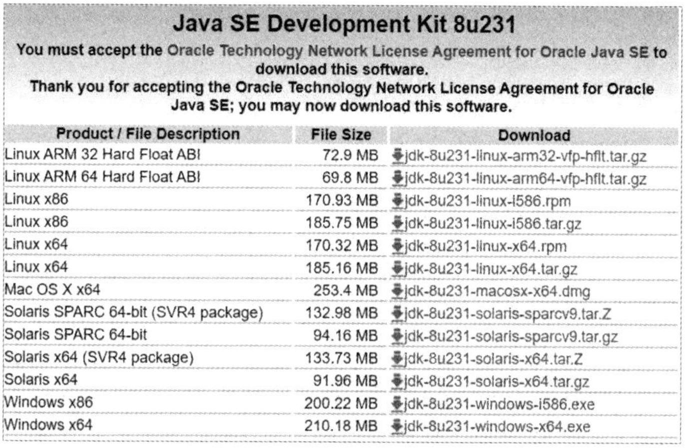

图1.1　Java SE Development Kit 8u231各平台安装包

根据提示，一步步安装 JDK。选择 JDK 的安装位置，如图 1.2 所示。

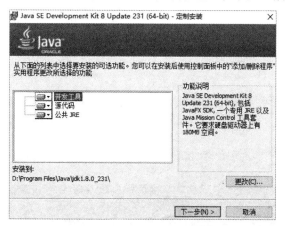

图1.2　选择JDK的安装位置

1　网址参见链接 1。

执行安装过程，待 JDK 安装完成后会安装 JRE，提示选择 JRE 的安装位置，如图 1.3 所示。

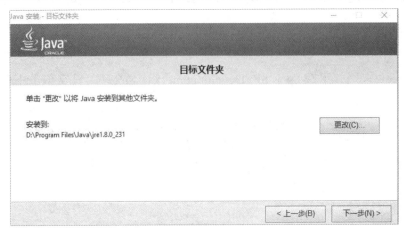

图1.3　安装过程

看到图 1.4，表明 Java 安装好了。

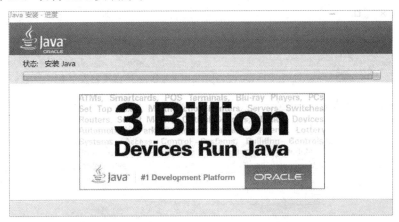

图1.4　安装成功

安装好后，打开 DOS 窗口，输入 java –version 命令，如果出现如图 1.5 所示的提示就代表 JDK 安装成功了。

```
C:\Users\10138>java -version
java version "1.8.0_231"
Java(TM) SE Runtime Environment (build 1.8.0_231-b11)
Java HotSpot(TM) 64-Bit Server VM (build 25.231-b11, mixed mode)
```

图1.5　Java版本号

在JetBrains网站[1]下载IntelliJ IDEA 2019.2.4，如图 1.6 所示。

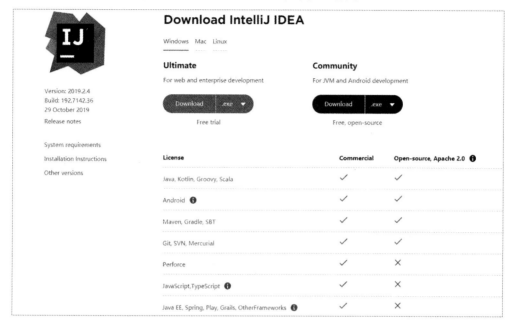

图1.6　IntelliJ IDEA 2019.2.4的下载页面

根据提示进行安装，选择安装位置，如图 1.7 所示。

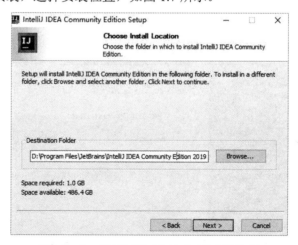

图1.7　IntelliJ IDEA 2019.2.4的安装位置

1　网址参见链接2。

然后执行安装过程，提取文件，如图1.8所示。

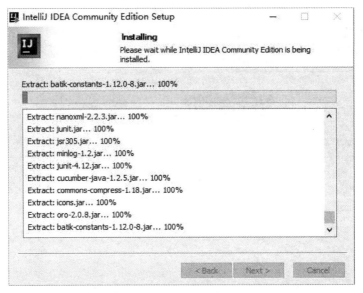

图1.8　IntelliJ IDEA 2019.2.4的安装过程

当出现如图1.9所示的界面时表示安装完成。

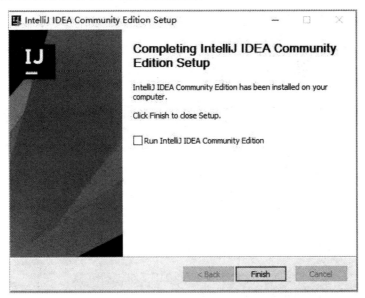

图1.9　IntelliJ IDEA 2019.2.4安装成功

1.3 在 Ubuntu 环境中搭建 Kotlin 开发环境

在如图 1.1 所示的界面和 IntelliJ IDEA 的网站下载相应的 Linux 平台的安装文件：jdk-8u231- linux-x64.tar.gz 和 ideaIC-2019.2.4.tar.gz。

运行如下命令将 jdk-8u231-linux-x64.tar.gz 解压到/opt 目录，如图 1.10 所示：

```
tar -xvf jdk-8u231-linux-x64.tar.gz
```

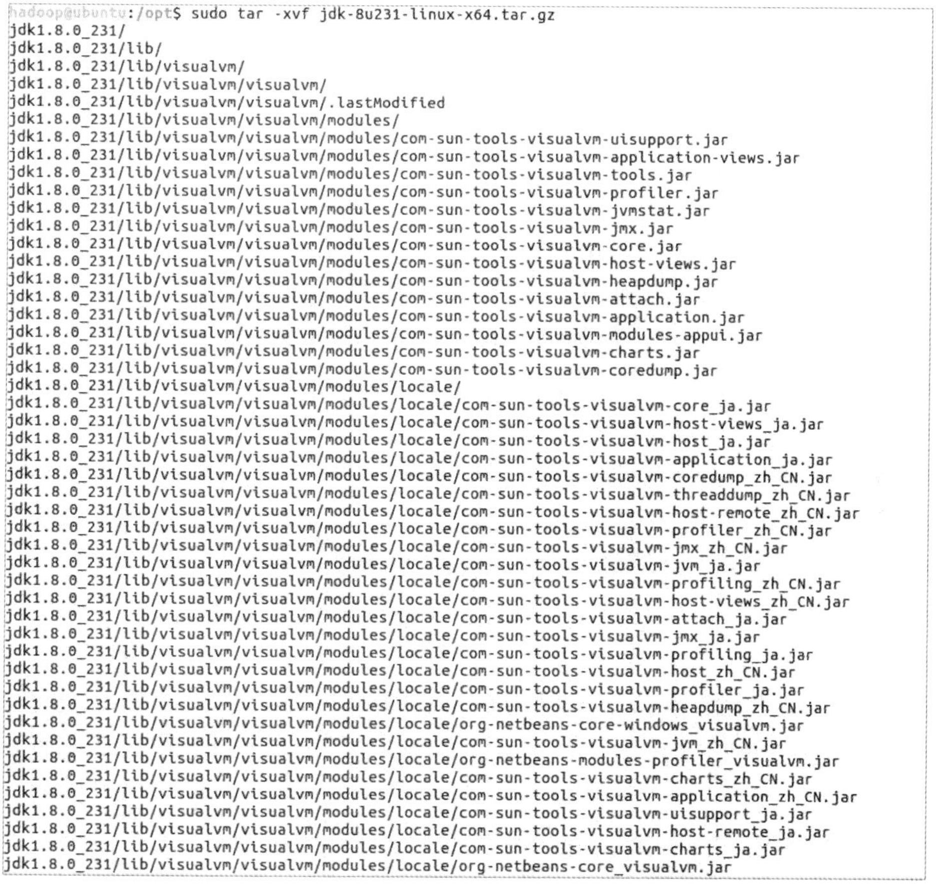

```
hadoop@ubuntu:/opt$ sudo tar -xvf jdk-8u231-linux-x64.tar.gz
jdk1.8.0_231/
jdk1.8.0_231/lib/
jdk1.8.0_231/lib/visualvm/
jdk1.8.0_231/lib/visualvm/visualvm/
jdk1.8.0_231/lib/visualvm/visualvm/.lastModified
jdk1.8.0_231/lib/visualvm/visualvm/modules/
jdk1.8.0_231/lib/visualvm/visualvm/modules/com-sun-tools-visualvm-uisupport.jar
jdk1.8.0_231/lib/visualvm/visualvm/modules/com-sun-tools-visualvm-application-views.jar
jdk1.8.0_231/lib/visualvm/visualvm/modules/com-sun-tools-visualvm-tools.jar
jdk1.8.0_231/lib/visualvm/visualvm/modules/com-sun-tools-visualvm-profiler.jar
jdk1.8.0_231/lib/visualvm/visualvm/modules/com-sun-tools-visualvm-jvmstat.jar
jdk1.8.0_231/lib/visualvm/visualvm/modules/com-sun-tools-visualvm-jmx.jar
jdk1.8.0_231/lib/visualvm/visualvm/modules/com-sun-tools-visualvm-core.jar
jdk1.8.0_231/lib/visualvm/visualvm/modules/com-sun-tools-visualvm-host-views.jar
jdk1.8.0_231/lib/visualvm/visualvm/modules/com-sun-tools-visualvm-heapdump.jar
jdk1.8.0_231/lib/visualvm/visualvm/modules/com-sun-tools-visualvm-attach.jar
jdk1.8.0_231/lib/visualvm/visualvm/modules/com-sun-tools-visualvm-application.jar
jdk1.8.0_231/lib/visualvm/visualvm/modules/com-sun-tools-visualvm-modules-appui.jar
jdk1.8.0_231/lib/visualvm/visualvm/modules/com-sun-tools-visualvm-charts.jar
jdk1.8.0_231/lib/visualvm/visualvm/modules/com-sun-tools-visualvm-coredump.jar
jdk1.8.0_231/lib/visualvm/visualvm/modules/locale/
jdk1.8.0_231/lib/visualvm/visualvm/modules/locale/com-sun-tools-visualvm-core_ja.jar
jdk1.8.0_231/lib/visualvm/visualvm/modules/locale/com-sun-tools-visualvm-host-views_ja.jar
jdk1.8.0_231/lib/visualvm/visualvm/modules/locale/com-sun-tools-visualvm-host_ja.jar
jdk1.8.0_231/lib/visualvm/visualvm/modules/locale/com-sun-tools-visualvm-application_ja.jar
jdk1.8.0_231/lib/visualvm/visualvm/modules/locale/com-sun-tools-visualvm-coredump_zh_CN.jar
jdk1.8.0_231/lib/visualvm/visualvm/modules/locale/com-sun-tools-visualvm-threaddump_zh_CN.jar
jdk1.8.0_231/lib/visualvm/visualvm/modules/locale/com-sun-tools-visualvm-host-remote_zh_CN.jar
jdk1.8.0_231/lib/visualvm/visualvm/modules/locale/com-sun-tools-visualvm-profiler_zh_CN.jar
jdk1.8.0_231/lib/visualvm/visualvm/modules/locale/com-sun-tools-visualvm-jmx_zh_CN.jar
jdk1.8.0_231/lib/visualvm/visualvm/modules/locale/com-sun-tools-visualvm-jvm_ja.jar
jdk1.8.0_231/lib/visualvm/visualvm/modules/locale/com-sun-tools-visualvm-profiling_zh_CN.jar
jdk1.8.0_231/lib/visualvm/visualvm/modules/locale/com-sun-tools-visualvm-host-views_zh_CN.jar
jdk1.8.0_231/lib/visualvm/visualvm/modules/locale/com-sun-tools-visualvm-attach_ja.jar
jdk1.8.0_231/lib/visualvm/visualvm/modules/locale/com-sun-tools-visualvm-jmx_ja.jar
jdk1.8.0_231/lib/visualvm/visualvm/modules/locale/com-sun-tools-visualvm-profiling_ja.jar
jdk1.8.0_231/lib/visualvm/visualvm/modules/locale/com-sun-tools-visualvm-host_zh_CN.jar
jdk1.8.0_231/lib/visualvm/visualvm/modules/locale/com-sun-tools-visualvm-profiler_ja.jar
jdk1.8.0_231/lib/visualvm/visualvm/modules/locale/com-sun-tools-visualvm-heapdump_zh_CN.jar
jdk1.8.0_231/lib/visualvm/visualvm/modules/locale/org-netbeans-core-windows_visualvm.jar
jdk1.8.0_231/lib/visualvm/visualvm/modules/locale/com-sun-tools-visualvm-jvm_zh_CN.jar
jdk1.8.0_231/lib/visualvm/visualvm/modules/locale/org-netbeans-modules-profiler_visualvm.jar
jdk1.8.0_231/lib/visualvm/visualvm/modules/locale/com-sun-tools-visualvm-charts_zh_CN.jar
jdk1.8.0_231/lib/visualvm/visualvm/modules/locale/com-sun-tools-visualvm-application_zh_CN.jar
jdk1.8.0_231/lib/visualvm/visualvm/modules/locale/com-sun-tools-visualvm-uisupport_ja.jar
jdk1.8.0_231/lib/visualvm/visualvm/modules/locale/com-sun-tools-visualvm-host-remote_ja.jar
jdk1.8.0_231/lib/visualvm/visualvm/modules/locale/com-sun-tools-visualvm-charts_ja.jar
jdk1.8.0_231/lib/visualvm/visualvm/modules/locale/org-netbeans-core_visualvm.jar
```

图1.10　jdk-8u231-linux-x64.tar.gz的安装过程

解压完会生成一个目录 jdk1.8.0_231/。然后在.bashrc 文件中配置环境变量，如下：

```
export JAVA_HOME=/opt/jdk1.8.0_231/
export JRE_HOME=${JAVA_HOME}/jre
```

```
export CLASSPATH=.:${JAVA_HOME}/lib:${JRE_HOME}/lib
export PATH=${JAVA_HOME}/bin:$PATH
```

执行 source .bashrc，使环境变量生效，然后运行 java -version，出现如图 1.11 所示的内容，表示 Java 安装成功。

```
hadoop@ubuntu:/opt$ java -version
java version "1.8.0_231"
Java(TM) SE Runtime Environment (build 1.8.0_231-b11)
Java HotSpot(TM) 64-Bit Server VM (build 25.231-b11, mixed mode)
```

图1.11　Java版本号

接下来安装 IntelliJ IDEA，将 ideaIC-2019.2.4.tar.gz 解压到/opt 目录，执行如下命令：

```
tar -xvf ideaIC-2019.2.4.tar.gz
```

安装过程如图 1.12 所示：

```
hadoop@ubuntu:/opt$ sudo tar -xvf ideaIC-2019.2.4.tar.gz
idea-IC-192.7142.36/LICENSE.txt
idea-IC-192.7142.36/NOTICE.txt
idea-IC-192.7142.36/bin/appletviewer.policy
idea-IC-192.7142.36/bin/idea.svg
idea-IC-192.7142.36/bin/log.xml
idea-IC-192.7142.36/build.txt
idea-IC-192.7142.36/lib/FastInfoset-1.2.15.jar
idea-IC-192.7142.36/lib/aapt-proto-jarjar.jar
idea-IC-192.7142.36/lib/annotations.jar
idea-IC-192.7142.36/lib/ant/CONTRIBUTORS
idea-IC-192.7142.36/lib/ant/INSTALL
idea-IC-192.7142.36/lib/ant/KEYS
idea-IC-192.7142.36/lib/ant/LICENSE
idea-IC-192.7142.36/lib/ant/NOTICE
idea-IC-192.7142.36/lib/ant/README
idea-IC-192.7142.36/lib/ant/WHATSNEW
idea-IC-192.7142.36/lib/ant/contributors.xml
idea-IC-192.7142.36/lib/ant/lib/README
idea-IC-192.7142.36/lib/ant/lib/ant-antlr.jar
idea-IC-192.7142.36/lib/ant/lib/ant-antlr.pom
idea-IC-192.7142.36/lib/ant/lib/ant-apache-bcel.jar
idea-IC-192.7142.36/lib/ant/lib/ant-apache-bcel.pom
idea-IC-192.7142.36/lib/ant/lib/ant-apache-bsf.jar
idea-IC-192.7142.36/lib/ant/lib/ant-apache-bsf.pom
idea-IC-192.7142.36/lib/ant/lib/ant-apache-log4j.jar
idea-IC-192.7142.36/lib/ant/lib/ant-apache-log4j.pom
idea-IC-192.7142.36/lib/ant/lib/ant-apache-oro.jar
idea-IC-192.7142.36/lib/ant/lib/ant-apache-oro.pom
idea-IC-192.7142.36/lib/ant/lib/ant-apache-regexp.jar
idea-IC-192.7142.36/lib/ant/lib/ant-apache-regexp.pom
idea-IC-192.7142.36/lib/ant/lib/ant-apache-resolver.jar
idea-IC-192.7142.36/lib/ant/lib/ant-apache-resolver.pom
idea-IC-192.7142.36/lib/ant/lib/ant-apache-xalan2.jar
idea-IC-192.7142.36/lib/ant/lib/ant-apache-xalan2.pom
idea-IC-192.7142.36/lib/ant/lib/ant-commons-logging.jar
idea-IC-192.7142.36/lib/ant/lib/ant-commons-logging.pom
idea-IC-192.7142.36/lib/ant/lib/ant-commons-net.jar
idea-IC-192.7142.36/lib/ant/lib/ant-commons-net.pom
idea-IC-192.7142.36/lib/ant/lib/ant-jai.jar
idea-IC-192.7142.36/lib/ant/lib/ant-jai.pom
idea-IC-192.7142.36/lib/ant/lib/ant-javamail.jar
idea-IC-192.7142.36/lib/ant/lib/ant-javamail.pom
idea-IC-192.7142.36/lib/ant/lib/ant-jdepend.jar
idea-IC-192.7142.36/lib/ant/lib/ant-jdepend.pom
idea-IC-192.7142.36/lib/ant/lib/ant-jmf.jar
idea-IC-192.7142.36/lib/ant/lib/ant-jmf.pom
idea-IC-192.7142.36/lib/ant/lib/ant-jsch.jar
```

图1.12　ideaIC-2019.2.4.tar.gz的安装过程

解压后会生成一个新目录 idea-IC-192.7142.36/。为了能够在桌面上启动 IntelliJ IDEA，需要创建快捷方式。在/usr/share/applications/目录下创建 intellij-idea.desktop，将如下内容复制到 intellij-idea.desktop 中：

```
1.  [Desktop Entry]
2.  Name=IntelliJ IDEA
3.  Exec=/opt/idea-IC-192.7142.36/bin/idea.sh
4.  Comment=IntelliJ IDEA
5.  Icon=/opt/idea-IC-192.7142.36/bin/idea.png
6.  Type=Application
7.  Terminal=false
8.  Encoding=UTF-8
```

保存后，执行 sudo chmod +x intellij-idea.desktop 命令。按 Windows 键，在搜索框中输入"int"，如图 1.13 所示。

图1.13　IntelliJ IDEA图标

将"IntelliJ IDEA"这个图标拖曳到桌面，这样可以方便快速启动 IntelliJ IDEA，如图 1.14 所示。

双击图标，启动 IntelliJ IDEA，可以看到版本号，表明 IntelliJ IDEA 安装成功，如图 1.15 所示。

图1.14　IntelliJ IDEA的快捷方式图标　　图1.15　IntelliJ IDE安装成功

1.4　在 macOS 环境中搭建 Kotlin 开发环境

在如图 1.1 所示的界面和 IntelliJ IDEA 的网站下载相应的 macOS 平台的安装文件：jdk-8u231-macosx-x64.dmg、ideaIC-2019.2.4.dmg。

双击 jdk-8u231-macosx-x64.dmg 安装包，打开如图 1.16 所示的窗口，按照系统提示进行安装。

图1.16　jdk-8u231-macosx-x64.dmg的安装过程

然后就可以配置系统的 Java 环境变量了。打开终端，进入当前用户的 home 目录，打开.bash_profile 并编辑：在文件的末尾加入下面这行语句并保存。

export JAVA_HOME=/Library/Java/JavaVirtualMachines/jdk1.8.0_211.jdk/Contents/Home

双击 ideaIC-2019.2.4.dmg 安装包，待进度条读完之后，拖曳 IDEA 图标到 Applications，如图 1.17 所示。

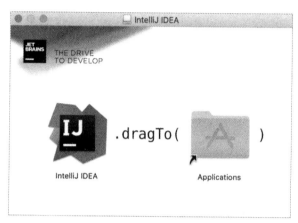

图1.17　安装ideaIC-2019.2.4.dmg

1.5　第一个 Kotlin 程序

本书使用 Windows 平台进行介绍。下面创建第一个 Kotlin 程序。首先打开 IntelliJ IDEA，新建一个包 io.kang.chapter01，然后新建一个 FirstKotlinApp.kt 文件。Kotlin 程序的目录如图 1.18 所示。

图1.18　Kotlin程序的目录

Kotlin 程序代码如下：

```
1. package io.kang.chapter01          // 包名
```

```
2.  fun main() {                          // 主函数
3.      println("Hello, World!")          // 打印"Hello, World!"
4.  }
```

首先，定义一个包。然后，定义程序的 main 函数，这是程序的入口。Kotlin 1.3 版本定义 main() 函数时可以不带任何参数。main 函数没有指定返回类型，不返回任何值。最后是方法体，println 方法输出一个字符串。

运行这个程序，可以输出"Hello，World!"，如图 1.19 所示。

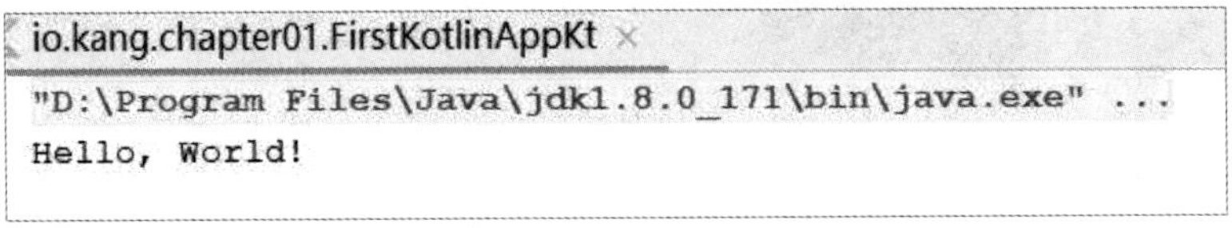

```
io.kang.chapter01.FirstKotlinAppKt
"D:\Program Files\Java\jdk1.8.0_171\bin\java.exe" ...
Hello, World!
```

图1.19　Kotlin程序运行结果

1.6　小结

本章首先以一个查找年龄最大的 Animal 对象的例子引入 Kotlin 语言，直观地展示了 Kotlin 语言的特性——它是静态类型语言、支持函数式编程和面向对象编程、具有很强的实用性、语法简洁、安全性高、和 Java 语言互操作性很好。

其次，本章图文并茂地介绍了在三大主流平台上安装 Java 1.8、IntelliJ IDEA 及 Kotlin 环境的过程，为后续章节编写 Kotlin 代码奠定了基础。

最后，本章用 IntelliJ IDEA 编写了一个 Kotlin 程序，以让大家对 Kotlin 有初步的了解。第 2 章将介绍函数式编程。

第 2 章

函数式编程介绍

本章主要介绍函数式编程的概念、特点。函数式编程具有不变性、惰性求值、引用透明、无副作用等特点。本章将介绍 Java、Kotlin、Scala 语言的特点，并对它们的差异进行比较。

2.1 初识函数式编程

函数式编程以纯函数的方式编写代码。纯函数指的是一个函数在程序的执行过程中除了根据输入参数给出运算结果外不受其他因素影响。纯函数的核心目的是编写无副作用的代码，它有很多特性，包括不变性、惰性求值等。

函数式编程更加强调程序执行的结果而非执行的过程，倡导利用若干简单的执行单元让计算结果不断演进，逐层推导复杂的运算，而不是设计一个复杂的执行过程。下面通过两个例子来比较一下命令式编程和函数式编程。

对一个数组进行累加。命令式编程逐个遍历元素，进行累加，代码如下：

```
1. public class AddDemo {
2.     // 声明数组
3.     public final static List<Integer> nums = Arrays.asList(0, 1, 2, 3, 4, 5, 6, 7, 8, 10);
```

```java
4.    // 加和方法
5.    public static Integer sum(List<Integer> nums) {
6.        int result = 0;
7.        for (Integer num : nums) {
8.            result += num;
9.        }
10.       return result;
11.   }
12.   // 主函数
13.   public static void main(String[] args) {
14.       System.out.println(AddDemo.sum(nums));
15.   }
16. }
```

使用函数式编程,设定初始值和运算方法(累加),遍历集合,即可获得累加和。函数式编程更加关注运算的结果:

```kotlin
1. fun main() {
2.     // 声明数组
3.     val nums = listOf(0, 1, 2, 3, 4, 5, 6, 7, 8, 10)
4.     // 打印数组元素的和
5.     println(nums.fold(0, { acc, i ->  acc + i}))
6. }
```

求斐波那契数列,命令式编程的代码如下,按照斐波那契数列的定义一步步计算结果:

```java
1. public class FibonacciDemo {
2.     // 斐波那契数列方法
3.     public static int fibonacci(int number) {
4.         if (number == 1) {
5.             return 1;
6.         }
7.         if(number == 2) {
8.             return 2;
9.         }
10.        int a = 1;
11.        int b = 2;
12.        for(int cnt = 3; cnt <= number; cnt++) {
```

```
13.         int c = a + b;
14.         a = b;
15.         b = c;
16.     }
17.     return b;
18. }
19. // 主函数
20. public static void main(String[] args) {
21.     System.out.println(FibonacciDemo.fibonacci(5));
22. }
23. }
```

求斐波那契数列，函数式编程的代码如下所示，使用函数式编程将每一步拆解为一个函数，通过函数的组合得到最终结果：

```
1.  // 斐波那契数列函数
2.  fun fibonacci(num: Int): Int {
3.      return Stream.iterate(listOf(1, 1), {
4.              t -> listOf(t[1], t[0] + t[1])
5.          })
6.          .limit(num.toLong())
7.          .map { n -> n[1] }
8.          .toList()[num - 1]
9.  }
10. // 主函数
11. fun main() {
12.     print(fibonacci(8))
13. }
```

2.2 函数式编程的特点

函数式编程具有如下特点。

函数是一等公民：函数与其他数据类型一样，处于平等地位，不但可以将函数作为参数、返回值使用，而且可以用函数名进行引用，甚至可以匿名调用。下面的例子展示了函数作为参数、返回值、匿名函数的用法。

```kotlin
1.  fun main() {
2.      val arr = listOf(1, 2, 3, 4)
3.      // 作为参数
4.      println(arr.reduce { acc, i -> acc + i })
5.      // 作为返回值
6.      fun operate(num: Int): (Int) -> Int {
7.          return {num -> num * 2}
8.      }
9.      // 匿名函数
10.     fun(x: Int, y: Int) = x + y
11. }
```

引用透明，无副作用：函数的运行不依赖外部变量或"状态"，只依赖输入的参数，任何时候只要参数相同，引用函数所得到的返回值总是相同的。副作用是指有超出函数控制的操作，比如在执行过程中操作文件系统、数据库等外部资源。没有副作用使得函数式编程各个独立部分的执行顺序可以被随意打乱，多个线程之间不共享状态，不会造成资源争用，不需要用锁来保护可变状态，不会出现死锁，这样可以更好地进行无锁的并发操作。

在下面的例子中，xs 是不可变对象，多次调用 xs.slice 方法，入参相同，返回结果相同。xs1 是可变对象，多次调用 xs1.add 方法，会改变 xs1 的值，产生副作用。

```kotlin
1.  fun main() {
2.      val xs = listOf(1, 2, 3, 4, 5)
3.      // 多次调用 xs.slice 方法的返回结果一致
4.      println(xs.slice(IntRange(0, 3)))
5.      println(xs.slice(IntRange(0, 3)))
6.      println(xs.slice(IntRange(0, 3)))
7.
8.      var xs1 = mutableListOf(1, 2, 3)
9.      // 多次调用 xs1.add 方法的返回结果不一样
10.     xs1.add(4)
11.     println(xs1)
12.     xs1.add(5)
13.     println(xs1)
14.     xs1.add(6)
15.     println(xs1)
16. }
```

高阶函数：可以将函数作为参数进行传递，也可以在函数内部声明一个函数，那么外层的函数就被称作高阶函数。

```
1.  // 函数 add 作为参数传递
2.  fun test(a: Int , b: Int, add: (num1 : Int , num2 : Int) -> Int) : Int{
3.      return add(a, b)
4.  }
5.  fun main() {
6.      // 调用 test 函数
7.      println(test(10, 11) { num1: Int, num2: Int -> num1 + num2 })
8.  }
```

柯里化（Currying）：把接收多个参数的函数变换成接收单一参数的函数，返回新函数，该函数接收余下的参数并且返回结果。柯里化的好处是减少了函数的参数个数，并且模块化了每一步计算，与设计模式中的适配器模式（将一个接口转换为另一个接口）类似。并且柯里化的应用之一——惰性求值，也是函数式编程的一个重要特性。惰性求值在将表达式赋值给变量（或称作绑定）时并不计算表达式的值，而是在变量第一次被使用时才进行计算。

下面的例子展示了柯里化、惰性求值。add(1)输入一个参数，返回一个函数；add(1)(2)，输入两个参数，返回一个函数；add(1)(2)(3)输入三个参数，返回一个结果。惰性求值在调用时才进行初始化，println(y)方法不会初始化 y，println(y.values)方法才会初始化 y。

```
1.  fun main() {
2.      // 定义一个柯里化函数 add,
3.      // 接收参数 a, 返回一个函数
4.      // 接收参数 b, 返回一个函数
5.      // 接收参数 c, 返回一个结果
6.      fun add(a: Int): (b: Int) -> (c: Int) -> Int {
7.          return {b -> {c -> a + b * c } }
8.      }
9.      // 调用函数 add
10.     println(add(1)(2)(3))
11.     // 懒加载
12.     val y = lazy {
13.         println("y")
14.         13
```

```
15.     }
16.     // 此时，13 没有实例化
17.     println(y)
18.     // 此时，13 实例化
19.     println(y.value)
20. }
```

2.3 Scala、Kotlin、Java 的对比

　　Scala、Kotlin、Java 都是以 Java 虚拟机（JVM）为目标运行环境的语言。Scala、Kotlin 运行于 JVM 上，所以 Scala、Kotlin 可以访问 Java 类库，能够与 Java 框架进行互操作。Scala、Kotlin 将面向对象和函数式编程有机结合在一起，既有动态语言的灵活性和简洁性，也有静态语言的类型检查功能来保证安全和提高执行效率。

　　Scala 是一门多范式的编程语言。Scala 具有面向对象的特性，每个值都是对象，对象的数据类型及行为由类和特性描述。类抽象机制的扩展有两种途径：一种途径是子类继承，另一种途径是灵活的混入机制。这两种途径能避免多重继承的种种问题。Scala 是一种函数式语言，其函数也能当成值来使用。Scala 提供了轻量级的语法，用于定义匿名函数，支持高阶函数，允许嵌套多层函数，并支持柯里化。可以利用 Scala 的模式匹配编写类似正则表达式的代码处理 XML 数据。Scala 具备类型系统，通过编译时检查，可保证代码的安全性和一致性。类型系统支持泛型类、协变和逆变、标注、类型参数的上下限约束、把类别和抽象类型作为对象成员、复合类型、引用自己时显式地指定类型、视图、多态方法、扩展性等。Scala 使用 Actor 作为其并发模型，Actor 是类似线程的实体，通过邮箱收发消息。Actor 可以复用线程，因此在程序中可以使用数百万个 Actor，而线程只能创建数千个。

　　Kotlin 语法和 Java 很像，容易上手，推荐以循序渐进的方式开发项目：允许项目中同时存在 Java 和 Kotlin 代码文件，允许 Java 与 Kotlin 互相调用。这使得开发者可以很方便地在已有项目中引入 Kotlin。Kotlin 代码量少且代码末尾没有分号。Kotlin 是空安全的，在编译时期就处理了各种 null 的情况，避免了执行时异常。Kotlin 可扩展函数，可以扩展任意类的更多的特性。Kotlin 也是函数式的，可使用 Lambda 表达式来更方便地解决问题。Kotlin 具有高度互操作性，可以继续使用所有有用 Java 编写的代码和库，甚至可以在一个项目中使用 Kotlin 和 Java 两种语言混合编程。

　　Java 是企业级的开发语言，很多互联网公司会使用 Java 开发后台应用，目前主要使用 Java 7 和 Java 8 进行开发。Java 是面向对象的、分布式的、健壮的、安全的，近些年也增

加了一些函数式语言的特性。Java 8 中引入了 Lambda 表达式、Stream，提供了类似 Scala、Kotlin 的函数式运算。Java 8 使用 Optional 避免空指针；Java 11 中引入了本地变量类型推断，增强了 Stream，增强了 Optional。

Kotlin 是更好的 Java，Kotlin 是一种实用的语言，旨在提高 Java 开发人员的工作效率，在 Java 模型之上有一些 Scala 特性。它增加了 Java 程序员想要的功能，如 Lambda 表达式和基本功能。虽然 Kotlin 为函数式编程提供了一些支持，但它确实可以实现更简单的过程式编程或命令式编程。

Scala 比 Java 更强大，旨在完成 Java 无法做到的事情。Scala 为高级函数编程提供了很好的支持，这增加了复杂性，使 Scala 成为一门比较难学习的语言。Scala 中有多种编程风格，可能会导致混乱或为每种需求提供最佳风格，这会导致更高的开发成本。

Kotlin 可以很方便地融入 Java 已有的生态，如 Spring Boot、Spring Cloud、Dubbo 等。Spring 5 引入了对 Kotlin 的支持。Scala 有自己的生态，目前其比较流行的 Web 框架是 Lift 框架和 Play 框架，微服务框架有 Akka，响应式微服务框架为 Lagom Framework。Scala 的框架有一定的学习门槛和难度，对普通 Java 开发者来讲不太容易掌握。

基于以上比较，Kotlin 和 Java 良好的互操作性使得 Java 开发者可以很快地掌握使用 Kotlin 开发微服务应用的方法，这也是本书的目标。

2.4 小结

本章介绍了函数式编程及其特点，函数是一等公民，函数式编程引用透明、无副作用，支持高阶函数，可柯里化并进行惰性求值等。本章还比较了 Scala、Kotlin、Java 三种编程语言的差异，Scala 具有隐式转换、隐式参数等很多 Kotlin 没有的特性；Spring 5 对 Kotlin 支持较好，可以利用 Java 的生态组件；Java 新增了 Lambda 表达式、类型推断、不可变集合等特性，逐渐向函数式编程靠拢。

第 3 章
Kotlin 的语法

本章介绍 Kotlin 的语法，包括基础语法：基本数据类型、包名和引用、流程控制、返回和跳转；类：接口、特殊类、泛型、委托；函数和 Lambda 表达式；集合：集合类型、集合操作；协程：协程基础知识及进阶。通过本章的介绍，读者对 Kotlin 的语法会有直观且快速的了解。

3.1 基础语法

本节主要介绍 Kotlin 使用的基本数据类型、包名和引用、流程控制、返回和跳转等基础语法。

3.1.1 基本数据类型

Kotlin 使用的基本数据类型包括：数值型、字符型、布尔型、数组和字符串。

数值型分为整数和浮点数。整数有 8 种类型，如表 3.1 所示。

表 3.1 Kotlin 中整数的类型

类型	位数（bit）	最小值	最大值
Byte	8	−128	127
Short	16	−32,768	32,767

续表

类型	位数（bit）	最小值	最大值
Int	32	$-2,147,483,648\ (-2^{31})$	$2,147,483,647\ (2^{31}-1)$
Long	64	$-9,223,372,036,854,775,808\ (-2^{63})$	$9,223,372,036,854,775,807\ (2^{63}-1)$
UByte	8	0	255
UShort	16	0	65,535
UInt	32	0	$2^{32}-1$
ULong	64	0	$2^{64}-1$

浮点数有 Float 和 Double 两种类型，如表 3.2 所示。

表 3.2 Kotlin 中浮点数的类型

类型	位数（bit）	有效位数	指数位数	小数位数
Float	32	24	8	6 或 7
Double	64	53	11	15 或 16

Kotlin 不允许隐式转换类型。例如，方法 printDouble() 的参数是 Double 类型的，调用该方法时，参数必须是 Double 类型的。每种类型都提供了如下显式转换方法。

- toByte(): Byte
- toShort(): Short
- toInt(): Int
- toLong(): Long
- toFloat(): Float
- toDouble(): Double
- toChar(): Char

```
1.  fun main() {
2.      // 打印数值
3.      fun printDouble(d: Double) { print(d) }
4.      // 声明变量
5.      val i = 1
6.      val d = 1.1
7.      val f = 1.1f
8.
9.      printDouble(d)
10.     printDouble(i) // 错误：类型不匹配
```

```
11.     printDouble(f) // 错误：类型不匹配
12. }
```

注意：Long 类型的数字的结尾有 L，如 123L。十六进制数以 0X 开头，如 0X0F。二进制数以 0b 开头，如 0b00001011。Float 类型的数字的结尾有 f 或 F。无符号数值，以 u 或 U 结尾。无符号长整数以 uL 或 UL 结尾。

数值也支持用"_"增强可读性，如下面的代码所示：

```
1. // 数值使用"_"增强可读性
2. val oneMillion = 1_000_000
3. val creditCardNumber = 1234_5678_9012_3456L
4. val socialSecurityNumber = 999_99_9999L
5. val hexBytes = 0xFF_EC_DE_5E
6. val bytes = 0b11010010_01101001_10010100_10010010
```

注意：这些数值类型可为空时（例如，Int?），数值是包装类，其他情况下的数值是基本数据类型。包装类的值相等，但是内存地址不同，如下面的代码所示：

```
1. // a 是包装类，==比较的是值是否相等，===比较的是地址是否相等
2. val a: Int = 10000
3. println(a === a) // 打印 'true'
4. val boxedA: Int? = a
5. val anotherBoxedA: Int? = a
6. println(boxedA === anotherBoxedA) // !!!打印 'false'!!!
7. println(boxedA == anotherBoxedA) //打印 'true'
```

数值类型支持使用==、!=、>、<、>=、<=进行范围判断。此外，Int 和 Long 类型的数值支持如下位运算。

- shl(bits)——有符号数值左移
- shr(bits)——有符号数值右移
- ushr(bits)——无符号数值右移
- and(bits)——按位与
- or(bits)——按位或
- xor(bits)——按位异或
- inv()——按位取反

- // in 的用法，表示范围
- val x = 1.01
- val isInRange = x in 1.0..2.0
- val isNotInRange = x !in 0.0..1.0
- println(isInRange) //打印 'true'
- println(isNotInRange) //打印 'true'

字符类型的关键字是 Char。字符类型用单引号，如'a'。字符类型不能直接转换为数值类型，需要进行显式转换。此外，如果字符类型可为空，其是包装类，而不是基本数据类型。

布尔类型的关键字是 Boolean，只有两个取值 true 和 false。如果布尔类型可为空，其也是包装类。布尔类型支持&&、||、!三种运算符。

数组类型的关键字是 Array，可以用 arrayOf()创建一个数组，例如，arrayOf(1,2,3)表示数组[1,2,3]。数组有 get()和 set()方法，可以用[]替代这两个方法，举例如下：

```
1. // 数组类型举例
2. val arrayTemp = arrayOf(1,2,3)
3. arrayTemp[0] = 0
4. println(arrayTemp[0]) //打印 '0'
5. println(arrayTemp.size) //打印 '3'
```

Kotlin 提供 byteArray、shortArray 和 intArray 等用于创建不同类型的数组。

```
1. // 基本数据类型数组举例
2. val intArray: IntArray = intArrayOf(1, 2, 3)
3. val byteArray: ByteArray = byteArrayOf(1, 2, 3)
4. val shortArray: ShortArray = shortArrayOf(1, 2, 3)
5. val longArray: LongArray = longArrayOf(1, 2, 3)
6. val charArray: CharArray = charArrayOf('1', '2', '3')
7. val floatArray: FloatArray = floatArrayOf(1.0f, 2.0f, 3.0f)
8. val doubleArray: DoubleArray = doubleArrayOf(1.0, 2.0, 3.0)
9. val booleanArray: BooleanArray = booleanArrayOf(true, false, true)
```

String 表示字符串。字符串是不可变对象，可以用[i]遍历每一个字符。

```
1. // 遍历字符串
2. val str = "hello"
3. for (c in str) {
```

```
4.     println(c)
5. }
```

可以用"+"拼接字符串：

```
1. // 拼接字符串
2. val s = "abc" + 1
3. println(s + "def")
```

字符串可以由"""分割，包含换行符：

```
1. // 使用"""初始化字符串，包含换行符
2. val text = """
3.     for (c in "foo")
4.     print(c)
5. """
```

在上面这个例子中，每行都输出了很多空格，可以用 trimMargin() 方法去除空格，每行都有一个"|"前缀，也可以自定义前缀，如">"，调用 trimMargin(">") 即可：

```
1. // 使用 trimMargin() 方法去除每行字符串前的空格
2. val text1 = """
3. |Tell me and I forget.
4. |Teach me and I remember.
5. |Involve me and I learn.
6. |(Benjamin Franklin)
7. """.trimMargin()
```

Kotlin 的 String 支持模板，在字符串中使用"$"可嵌入模板：

```
1. // $支持字符串模板
2. val strTemplate = "abc"
3. // 打印结果："abc.length is 3"
4. println("$strTemplate.length is ${strTemplate.length}")
```

如果需要单独使用$，可以参考如下代码来实现：

```
1. //字符串模板单独使用$
2. val price = """
```

```
3.    |${'$'}9.99
4.    """.trimMargin()
5. println(price) //prints $9.99
```

3.1.2 包名和引用

Kotlin 的源码中一般都有一个包名，所有的类和方法等都在这个包下：

```
1. package io.kang.chapter03.basicsyntax
2. // 方法
3. fun printMessage() { /*...*/ }
4. // 类
5. class Message { /*...*/ }
```

在上面这个例子中，printMessage 的全名是 io.kang.chapter03.basicsyntax.printMessage，Message 的全名是 io.kang.chapter03.basicsyntax.Message。

每一个 Kotlin 文件在 JVM 平台中会默认引入如下包：

- kotlin.*
- kotlin.annotation.*
- kotlin.collections.*
- kotlin.comparisons.*（从 1.1 版本开始）
- kotlin.io.*
- kotlin.ranges.*
- kotlin.sequences.*
- kotlin.text.*
- java.lang.*
- kotlin.jvm.*

我们可以引入一个类，也可以引入包中的所有内容，为了避免命名冲突，还可以给引用起别名：

```
1. // 引用包
2. import io.kang.chapter03.basicsyntax.Message
3. import io.kang.chapter03.basicsyntax.*
4. import io.kang.chapter03.basicsyntax.Message as Message1
```

3.1.3 流程控制

Kotlin 的流程控制语句有 if、when、for、while。if 的分支可以是代码块,最后的表达式作为该块的值。

```
1.  // 传统用法
2.  var maxVal = 1
3.  if (1 < 2) maxVal = 2
4.
5.  // 使用 else
6.  if (1 > 2) {
7.      maxVal = 1
8.  } else {
9.      maxVal = 2
10. }
11. // maxVal 作为表达式
12. maxVal = if (1 > 2) 1 else 2
13. maxVal = if (1 > 2) {
14.     print("Choose 1")
15.     1
16. } else {
17.     print("Choose 2")
18.     2
19. }
```

when 语句类似于 switch,when 将它的参数与所有的分支条件顺序比较,直到某个分支满足条件。when 既可以被当作表达式使用,也可以被当作语句使用。

```
1.  // when 用法举例
2.  when (0) {
3.      is Int -> println("x is Int")
4.      else -> println("other type")
5.  }
```

for 语句可以对任何提供迭代器(iterator)的对象进行遍历。

```
1.  // for 升序遍历范围
2.  for (i in 1..3) {
```

```
3.     println(i)
4.  }
5.  // for 降序遍历范围
6.  for (i in 6 downTo 0 step 2) {
7.     println(i)
8.  }
9.  // for 遍历数组
10. val array = arrayOf(1,2,3)
11. for (i in array.indices) {
12.    println(array[i])
13. }
14. // for 解析数组的下标和元素
15. for ((index, value) in array.withIndex()) {
16.    println("the element at $index is $value")
17. }
```

While、do…while 在 Kotlin 中的用法和在 Java 中相同。

3.1.4 返回和跳转

Kotlin 有以下三种跳转表达式：

return——默认从最直接包围它的函数或者匿名函数返回。

break——终止最直接包围它的循环。

continue——继续下一次最直接包围它的循环。

在 Kotlin 中，任何表达式都可以用标签（label）来标记。标签的格式为在标识符后跟 @符号，例如 abc@。可以用标签限制 return、break、continue，举例如下：

```
1.  // return 举例
2.  fun foo() {
3.     listOf(1, 2, 3, 4, 5).forEach {
4.         if (it == 3) return // 直接返回到 foo()的调用处
5.         print(it)
6.     }
7.     println("this point is unreachable")
8.  }
9.  // 使用 break 跳转到 loop@
```

```
10. fun foo1() {
11.     loop@ for (i in 1..5) {
12.         for (j in 1..5) {
13.             if (j == 2) break@loop
14.             println("$i---$j")
15.         }
16.     }
17. }
18. // 使用 continue 跳转到 loop@
19. fun foo2() {
20.     loop@ for (i in 1..5) {
21.         for (j in 1..5) {
22.             if (j == 2) continue@loop
23.             println("$i---$j")
24.         }
25.     }
26. }
```

3.2 类

本节介绍如何定义 Kotlin 中的类、属性、接口，以及特殊类、泛型、委托等语法的概念和用法。

3.2.1 类、属性、接口

在 Kotlin 中，使用关键字 class 声明类。类声明由类名、类头（指定其类型参数、主构造函数等）及由花括号包围的类体构成，其中类体中包括构造函数、初始化块、函数、属性、嵌套类与内部类、对象声明。类头与类体都是可选的；如果一个类没有类体，可以省略花括号。Kotlin 中的一个类可以有一个主构造函数及一个或多个次构造函数。如果主构造函数没有任何注解或者可见性修饰符，可以省略 constructor 关键字。主构造函数不能包含任何代码。初始化代码可以放到以 init 关键字作为前缀的初始化块（initializer block）中。

主构造函数中声明的属性可以是可变的（var）或只读的（val），构造函数的属性可以有默认值。类也可以声明前缀为 constructor 的次构造函数。如果类有一个主构造函数，每个次构造函数需要委托给主构造函数,可以直接委托或者通过其他次构造函数进行间接委托。委托到同一个类的另一个构造函数用 this 关键字即可。所有初始化块中的代码都会在次构造函数体之前执行。即使该类没有主构造函数,这种委托仍会隐式发生,并且仍会执行初始化块。

```kotlin
1.  package io.kang.chapter03.basicsyntax
2.  // 定义 Person 类
3.  class Person(val name: String, var age: Int = 0) {
4.      var children: MutableList<Person> = mutableListOf<Person>();
5.      // 初始化块
6.      init {
7.          println("name is $name, age is $age")
8.      }
9.      // 次构造函数
10.     constructor(name: String, age: Int, parent: Person) : this(name, age) {
11.         parent.children.add(this)
12.     }
13. }
14. fun main() {
15.     // 实例化一个 Person 对象,使用主构造函数
16.     val person = Person("yuan")
17.     // 实例化一个 Person 对象,使用次构造函数
18.     val parentPerson = Person("kang", 1, person)
19. }
```

要创建一个类的实例,可以像普通函数一样调用构造函数。Kotlin 中没有 new 关键字,Kotlin 中的所有类都有一个共同的超类 Any,其对没有超类型声明的类来说是默认超类。Any 有三个方法:equals()、hashCode() 与 toString(),因此,为所有 Kotlin 类都定义了这些方法。

如果类要被继承,需要用 open 关键字修饰。Kotlin 对可覆盖的成员(我们称之为开放)及覆盖后的成员需要进行显式修饰。如果派生类有一个主构造函数,那么其基类类型可以(并且必须)用基类的主构造函数参数就地初始化。如果派生类没有主构造函数,那么每个次构造函数必须使用 super 关键字初始化其基类类型,或委托给另一个构造函数做到这一点。

```
1.  // 定义基类 Base,open 为关键字,表示 Base 可以被继承
```

```kotlin
2.  open class Base(val p: Int) {
3.      // open 修饰的属性，方法可以被子类覆盖
4.      open val vertexCount: Int = 0
5.      open fun draw() {
6.          println("Base.draw()")
7.      }
8.      constructor(base: String, p: Int): this(p){
9.          println("base is $base")
10.     }
11. }
12. // 定义 Derived，它继承了 Base
13. class Derived: Base {
14.     override var vertexCount = 4
15.     constructor(base: String, p: Int): super(base, p)
16. 
17.     override fun draw() {
18.         println("Derived.draw()")
19.     }
20. }
21. fun main() {
22.     // 实例化 Derived 类
23.     val derived = Derived("kang", 1)
24.     derived.draw()
25. }
```

Derived.draw()函数必须加上 override 修饰符，如果没加，编译器将会报错。如果函数没有标注 open，那么子类中不允许定义相同签名的函数。将 open 修饰符添加到 final 类（即没有 open 的类）的成员上会不起作用。标记为 override 的成员本身是开放的，也就是说，它可以在子类中被覆盖。如果你想禁止再次被覆盖，可使用 final 关键字。

类及其中的某些成员可以被声明为 abstract。抽象成员在本类中可以不用实现，可以用一个抽象成员覆盖一个非抽象的开放成员。

Kotlin 类中的属性既可以用关键字 var 声明为可变的，也可以用关键字 val 声明为只读的。要使用一个属性，只要用名称引用它即可。声明一个属性的完整语法如下：

```
var <propertyName>[: <PropertyType>] [= <property_initializer>]
    [<getter>]
    [<setter>]
```

```
1.  var allByDefault: Int? // 错误：需要进行显式初始化，提供默认的 getter 和 setter 方法
2.  var initialized = 1 // 类型为 Int，提供默认的 getter 和 setter 方法
3.  val simple: Int? // 类型为 Int，提供默认的 getter 方法必须在构造函数中初始化
4.  val inferredType = 1 // 类型为 Int，提供默认的 getter 方法
```

可以自定义 setter、getter 方法，举例如下：

```
1.  class Demo(val aList: MutableList<String>) {
2.      // 定义 size 的 getter 和 setter
3.      var size: Int = 0
4.          get() = aList.size
5.          set(value) {
6.              field = value
7.          }
8.      // 定义 listToString 的 getter 和 setter
9.      var listToString: String
10.         get() = aList.toString()
11.         set(value) {
12.             aList.add(value)
13.         }
14. }
15. // 实例化 Demo 对象，调用属性的 getter 方法
16. val demo = Demo(MutableList<String>(10){"a"})
17. println(demo.size)
18. println(demo.listToString)
```

常量可以用 const 修饰。如果要推后初始化属性和变量，可以用 lateinit 关键字。

Kotlin 的接口可以既包含抽象方法的声明也包含具体实现。与抽象类不同的是，接口无法保存状态。它可以有属性但必须声明为抽象或提供访问器实现，可使用关键字 interface 来定义接口。

```
1.  // 定义接口 Named
2.  interface Named {
3.      val name: String
4.  }
```

```kotlin
5.  // 定义接口 People，继承 Named
6.  interface People : Named {
7.      val firstName: String
8.      val lastName: String
9.      // name 属性的默认实现
10.     override val name: String get() = "$firstName $lastName"
11. }
12. // Employee 实现 People 接口
13. class Employee(
14.     // 不必实现"name"，接口有默认实现
15.     override val firstName: String,
16.     override val lastName: String
17. ) : People
```

类、对象、接口、构造函数、方法、属性和它们的 setter 都可以有可见性修饰符。Kotlin 有四个可见性修饰符：private、protected、internal 和 public，默认可见性是 public。private 只在类内部（包含其所有成员）可见；protected 在子类中可见。能见到类声明的本模块内的任何客户端都可见其 internal 成员；能见到类声明的任何客户端都可见其 public 成员。如果你覆盖一个 protected 成员并且没有显式指定其可见性，那么该成员还会是 protected 的。

```kotlin
1. // 文件名：classes.kt
2. package io.kang.chapter03.basicsyntax
3.
4. private fun foo() { ... } // 在 classes.kt 内可见
5.
6. public var bar: Int = 5 // 该属性随处可见
7.     private set         // setter 只在 classes.kt 内可见
8.
9. internal val baz = 6    // 相同模块内可见
```

3.2.2 特殊类

Kotlin 中的特殊类有如下几种。

数据类：我们经常创建一些只保存数据的类，用 data 标记。数据类的主构造函数需要至少有一个参数；所有参数标记为 val 或 var；数据类不能是抽象、开放、密封或者内部的。如果生成的类需要含有一个无参的构造函数，则所有的属性必须指定默认值。

```
1.  // 定义数据类 User
2.  data class User(val name: String = "", val age: Int = 0)
3.  // 实例化 User
4.  val jack = User(name = "Jack", age = 1)
5.  val olderJack = jack.copy(age = 2)
6.  // 析构 jack 的 name、age 属性
7.  val (name, age) = jack
8.  println("$name, $age years of age") // 输出 "Jane, 35 years of age"
```

密封类：用来表示受限的类继承结构，即一个值为有限的几种类型且不能有任何其他类型。密封类的子类是包含状态的多个实例。密封类需要在类名前面添加 sealed 修饰符。所有子类都必须在与密封类自身相同的文件中声明。密封类是抽象的，其构造函数默认是 private 的。

```
1.  // 密封类 Expr
2.  sealed class Expr
3.  // Const、Sum 继承 Expr
4.  data class Const(val number: Double) : Expr()
5.  data class Sum(val e1: Expr, val e2: Expr) : Expr()
6.  // 使用 when 进行类型判断
7.  fun eval(expr: Expr): Double = when(expr) {
8.      is Const -> expr.number
9.      is Sum -> eval(expr.e1) + eval(expr.e2)
10.     // 不再需要 else 子句，因为我们已经覆盖了所有的情况
11. }
```

嵌套类：嵌套在其他类内部的类。标记为 inner 的嵌套类能够访问其外部类的成员。

枚举类：其最基本的用法是实现类型安全的枚举，每一个枚举都可以初始化。枚举常量还可以声明其带有相应的方法及覆盖了基类方法的匿名类。

```
1.  // 枚举类
2.  enum class ProtocolState {
3.      WAITING {
4.          override fun signal() = TALKING
5.      },
```

```
6.      TALKING {
7.          override fun signal() = WAITING
8.      };
9.
10.     abstract fun signal(): ProtocolState
11. }
```

3.2.3 泛型

Kotlin 中的类也可以有类型参数。Kotlin 的类型系统有声明处型变（declaration-site variance）与类型投影（type projection）。

```
1.  // 定义一个泛型类 Box
2.  class Box<T>(t: T) {
3.      var value = t
4.  }
5.  // 实例化时，设置类型为 Int
6.  val box: Box<Int> = Box<Int>(1)
```

Kotlin 用 out 关键字表示泛型是协变类型的。当一个类 C 的类型参数 T 被声明为 out 时，它就只能出现在 C 的成员的输出位置，这样，C<Base>可以安全地作为 C<Derived>的超类。类 C 在参数 T 上是协变的，或者说 T 是一个协变的类型参数。C 是 T 的生产者，而不是 T 的消费者。out 修饰符也被称为型变注解，并且由于它在类型参数声明处提供，所以我们称之为声明处型变。

```
1.  // 定义协变类型
2.  interface Source<out T> {
3.      fun nextT(): T
4.  }
5.
6.  fun demo(strs: Source<String>) {
7.      val objects: Source<Any> = strs // 这个没问题，因为 T 是一个 out 参数
8.  }
```

与 out 相反，Kotlin 又补充了一个型变注解：in。它使得一个类型参数逆变：只可以被消费而不可以被生产。

```kotlin
// 定义逆变类型
interface Comparable<in T> {
    operator fun compareTo(other: T): Int
}
fun demo(x: Comparable<Number>) {
    x.compareTo(1.0) // 1.0 拥有类型 Double，它是 Number 的子类型
    // 因此，我们可以将 x 赋给类型为 Comparable <Double> 的变量
    val y: Comparable<Double> = x // 正确！
}
```

将类型参数 T 声明为 out 非常方便，但是有些类实际上不能被限制为只返回 T。例如 Array，该类在 T 上既不能是协变的也不能是逆变的。

```kotlin
// Array 在 T 上既不能是协变的，也不能是逆变的
class Array<T>(val size: Int) {
    fun get(index: Int): T { }
    fun set(index: Int, value: T) { }
}
```

对于如下例子，无法将一个 Array<Int>数组复制到 Array<Any>数组。Array<T>在 T 上是不型变的。

```kotlin
// copy()函数
fun copy(from: Array<Any>, to: Array<Any>) {
    assert(from.size == to.size)
    for (i in from.indices)
        to[i] = from[i]
}
fun copyValue() {
    val ints: Array<Int> = arrayOf(1, 2, 3)
    val any = Array<Any>(3) { "" }
    copy(ints, any)
    println(any.asList())
    // 其类型为 Array<Int>，但此处期望是 Array<Any>
}
```

我们唯一要确保的是，copy()不会做任何坏事。如果想阻止它写到 from，可以用这样的语句：

```
fun copy(from: Array<out Any>, to: Array<Any>) { …… }
```

这就是类型投影。我们所说的 from 不仅仅是一个数组，还是一个受限制的（投影的）数组，其只可以调用返回类型为类型参数 T 的方法，如上，这意味着只能调用 get()。同理，也可以用 in 投影一个类型。可以传递一个 CharSequence 数组或一个 Object 数组给 fill()函数。

```
1.  // 类型投影
2.  fun fill(dest: Array<in String>, value: String) {
3.      for(i in dest.indices){
4.          dest[i] = value
5.      }
6.  }
7.  // 定义一个数组，元素类型是 Any
8.  val arrayNulls = arrayOfNulls<Any>(2)
9.  // 将"hello"字符串复制到 arrayNulls 数组
10. fill(arrayNulls, "hello")
11. println(arrayNulls.asList())
```

3.2.4 委托

委托模式已经被证明是实现继承的一个很好的替代方式。DelagateDerived 类可以通过将其所有公有成员委托给指定对象来实现一个接口 Delagate。DelagateDerived 的超类型列表中的 by 子句表示 b 将会在 DelagateDerived 内部存储，并且编译器将生成转发给 b 的所有 Delagate 的方法。编译器会使用 override 覆盖的实现而不是委托对象中的实现。以这种方式重写的成员不会在委托对象的成员中调用，委托对象的内部成员只能访问自己实现的接口成员。

```
1.  // 定义接口 Delagate
2.  interface Delagate {
3.      val message: String
4.      fun print()
5.  }
6.  // 定义接口 Delagate 的实现类
7.  class DelagateImpl(val x: Int) : Delagate {
8.      override val message = "BaseImpl: x = $x"
9.      override fun print() { println(message) }
```

```
10. }
11. // 委托类
12. class DelagateDerived(b: Delagate) : Delagate by b {
13.     // 在 b 的 print 实现中不会访问到这个属性
14.     override val message = "Message of Derived"
15. }
16. fun main() {
17.     val b = DelagateImpl(10)
18.     val derived = DelagateDerived(b)
19.     // 打印：BaseImpl: x = 10
20.     derived.print()
21.     // 打印：Message of Derived
22.     println(derived.message)
23. }
```

Kotlin 支持委托属性、延迟属性、可观察属性等。语法是：val/var <属性名>: <类型> by <表达式>。by 后面的表达式是委托类，这是因为属性对应的 get()（与 set()）会被委托给它的 getValue() 与 setValue() 方法。属性的委托不必实现任何接口，但是需要提供一个 getValue() 函数（与 setValue()——对于 var 属性来说）。例如：

```
1.  // Example 类
2.  class Example {
3.      // 委托属性
4.      var p: String by Delegate()
5.  }
6.  // 代理类
7.  class Delegate {
8.      operator fun getValue(thisRef: Any?, property: KProperty<*>): String {
9.          return "$thisRef, thank you for delegating '${property.name}' to me!"
10.     }
11.
12.     operator fun setValue(thisRef: Any?, property: KProperty<*>, value: String) {
13.         println("$value has been assigned to '${property.name}' in $thisRef.")
14.     }
15. }
16. val e = Example()
```

```
17. //打印: //io.kang.chapter03.basicsyntax.Example@73a8dfcc, thank you for delegating
    '//p' to me!
18. println(e.p)
19. //打印: //NEW has been assigned to 'p' in io.kang.chapter03.basicsyntax.Example@73a8
    //dfcc.
20. e.p = "NEW"
```

3.3　函数和 Lambda 表达式

本节介绍函数的定义、参数、返回类型和 Lambda 表达式的操作。

3.3.1　函数

Kotlin 用 fun 关键字声明函数，函数的参数必须有显式类型，函数参数可以有默认值，当省略相应的参数时使用默认参数值。覆盖方法总是使用与基类类型方法相同的默认参数值。当覆盖一个带有默认参数值的方法时，必须从签名中省略默认参数值。如果一个默认参数在一个无默认值的参数之前，那么该默认值只能通过使用命名参数调用该函数来使用。如果在默认参数之后的最后一个参数是 Lambda 表达式，那么它作为命名参数既可以在括号内传入，也可以在括号外传入。函数的参数（通常是最后一个）可以用 vararg 修饰符修饰。

如果一个函数不返回任何有用的值，那么它的返回类型是 Unit。Unit 返回类型声明也是可选的。当函数返回单个表达式时，可以省略花括号并且在 "=" 之后指定代码体。当返回值类型可由编译器推断时，显式声明返回类型是可选的。

标有 infix 关键字的函数也可以使用中缀表示法（忽略该函数的点与圆括号）调用。中缀函数必须满足以下要求。

- 它们必须是成员函数或扩展函数。
- 它们只能有一个参数。
- 其参数不得接收可变数量的参数且不能有默认值。
- 中缀函数的优先级低于算术操作符、类型转换及 rangeTo 操作符。

```
1. // 定义类 A
2. open class A {
3.     open fun foo(i: Int = 10) { /*...*/ }
4. }
```

```kotlin
5.  // B 继承自 A
6.  class B : A() {
7.      override fun foo(i: Int) { /*...*/ }   // 不能有默认值
8.  }
9.  // 函数 foo
10. fun foo(bar: Int = 0, baz: Int) { /*...*/ }
11. // 函数 foo
12. fun foo(bar: Int = 0, baz: Int = 1, qux: () -> Unit): Unit { /*...*/ }
13.
14. // 单表达式函数
15. fun double(x: Int) = x * 2
16. // 泛型函数
17. fun <T> asList(vararg ts: T): List<T> {
18.     val result = ArrayList<T>()
19.     for (t in ts) // ts 是一个数组
20.         result.add(t)
21.     return result
22. }
23.
24. fun main() {
25.     foo(baz = 1) // 使用默认值 bar = 0
26.
27.     foo(1) { println("hello") }     // 使用默认值 baz = 1 的函数 foo
28.     // 使用两个默认值 bar = 0、baz = 1 的函数 foo
29.     foo(qux = { println("hello") })
30.     foo { println("hello") }
31.     val a = arrayOf(1, 2, 3)
32.     // 可以用*将数组展开
33.     val list = asList(-1, 0, *a, 4)
```

Kotlin 中的函数可以在局部作用域中声明,作为成员函数、顶层函数及扩展函数。局部函数,即一个函数在另一个函数内部,局部函数可以访问外部函数(即闭包)的局部变量。

```kotlin
1. data class Vertex(val neighbors: List<Vertex>)
2. data class Graph(val vertices: List<Vertex>)
3. fun dfs(graph: Graph) {
```

```
4.      val visited = HashSet<Vertex>()
5.      // dfs 是局部函数
6.      fun dfs(current: Vertex) {
7.          if (!visited.add(current)) return
8.          for (v in current.neighbors)
9.              dfs(v)
10.     }
11.     dfs(graph.vertices[0])
12. }
```

成员函数是在类或对象内部定义的函数。函数可以有泛型参数，通过在函数名前使用尖括号指定。Kotlin 支持一种称为尾递归的函数式编程风格，允许将一些通常用循环写的算法改用递归函数来写，从而避免堆栈溢出的风险。当一个函数用 tailrec 修饰符修饰并满足所需的形式时，编译器会优化该递归代码，留下一个快速而高效的基于循环的版本。

```
1. val eps = 1E-10
2. // 尾递归函数
3. tailrec fun findFixPoint(x: Double = 1.0): Double
4.         = if (Math.abs(x - Math.cos(x)) < eps) x else findFixPoint(Math.cos(x))
```

内联函数可以提高运行时效率，用 inline 修饰。inline 修饰符影响函数本身和传给它的 Lambda 表达式，所有这些都将内联到调用处。

```
inline fun <T> lock(lock: Lock, body: () -> T): T { ... }
```

3.3.2　Lambda 表达式

Kotlin 中的函数是一等公民，这意味着它们可以存储在变量与数据结构中、作为参数传递给其他高阶函数及从其他高阶函数返回，可以像操作任何其他非函数值一样操作函数。

为促成这一点，作为静态类型编程语言的 Kotlin 使用一系列函数类型表示函数并提供一组特定的语言结构，例如，Lambda 表达式。

高阶函数是将函数用作参数或返回值的函数。fold 函数接收一个初始累积值与一个接合函数，并通过将当前累积值与每个集合元素连续接合起来代入累积值来构建返回值。为了调用 fold，需要传给它一个函数类型的实例作为参数，而在高阶函数调用处，广泛使用 Lambda 表达式。

```kotlin
1.  fun main() {
2.      val items = listOf(1, 2, 3, 4, 5)
3.      // Lambda 表达式是花括号括起来的代码块
4.      items.fold(0, {
5.          // 如果一个 Lambda 表达式有参数，前面是参数，后跟"->"
6.          acc: Int, i: Int ->
7.          print("acc = $acc, i = $i, ")
8.          val result = acc + i
9.          println("result = $result")
10.         // Lambda 表达式中的最后一个表达式是返回值
11.         result
12.     })
13.     // Lambda 表达式的参数类型是可选的，如果能够将其推断出来的话，不需要声明参数类型
14.     val joinedToString = items.fold("Elements:", { acc, i -> acc + " " + i })
15.     // 打印：Elements: 1 2 3 4 5
16.     println(joinedToString)
17.     // 函数引用也可以用于高阶函数调用
18.     val product = items.fold(1, Int::times)
19.     // 打印：120
20.     println(product)
21. }
```

Kotlin 使用类似(Int) -> String 的一系列函数类型来处理函数的声明。这些类型具有与函数签名相对应的特殊表示法，即它们的参数和返回值。

有几种方法可以获得函数类型的实例。

使用函数字面值的代码块，采用以下形式之一。

- **Lambda 表达式**：{ a, b -> a + b }。
- **匿名函数**：fun(s: String): Int { return s.toIntOrNull() ?: 0 }。

使用已有声明的可调用引用。

- **顶层、局部、成员、扩展函数**：::isOdd、String::toInt。
- **顶层、成员、扩展属性**：List<Int>::size。
- **构造函数**：::Regex。

函数类型的值可以通过其 invoke()操作符调用：f.invoke(x)或者直接用 f(x)。

```
1. val stringPlus: (String, String) -> String = String::plus
2. val intPlus: Int.(Int) -> Int = Int::plus
3. // 打印：<-->
4. println(stringPlus.invoke("<-", "->"))
5. // 打印：Hello, world!
6. println(stringPlus("Hello, ", "world!"))
7. // 打印：2
8. println(intPlus.invoke(1, 1))
9. // 打印：3
10. println(intPlus(1, 2))
11. // 打印：5
12. println(2.intPlus(3)) // 类扩展调用
```

一个 Lambda 表达式只有一个参数是很常见的。如果编译器自己可以识别出签名，也可以不用声明唯一的参数并忽略->。该参数会被隐式声明为 it：

```
ints.filter { it > 0 } // 这个字面值是"(it: Int) -> Boolean"类型的
```

如果 Lambda 表达式的参数未使用，那么可以用下画线取代其名称：

```
map.forEach { _, value -> println("$value!") }
```

3.4 集合

本节介绍 Kotlin 中的 List、Set、Map 集合及集合的操作，包括添加元素、查找元素、过滤元素、转换等。

3.4.1 集合概述

Kotlin 标准库提供了一整套用于管理集合的工具，集合通常包含相同类型的一些（数目也可以为零）对象。集合中的对象称为元素或条目。以下是 Kotlin 相关的集合类型。

- List 是一个有序集合，可通过索引（反映元素位置的整数）访问元素。元素可以在 List 中出现多次。List 有一组字，这些字的顺序很重要并且字可以重复。
- Set 是唯一元素的集合。它反映了集合的数学抽象：一组不重复的对象。一般来说，Set 中元素的顺序并不重要。

- Map（或者字典）是一组键值对。键是唯一的，每个键都刚好映射到一个值，值可以重复。Map 对存储对象之间的逻辑连接非常有用。

只读集合类型是型变的。这意味着，如果类 Rectangle 继承自 Shape，则可以在需要 List<Shape>的任何地方使用 List<Rectangle>。换句话说，集合类型与元素类型具有相同的类继承关系。Map 在值（value）类型上是型变的，但在键（key）类型上不是。反之，可变集合不是型变的；否则将导致运行时故障。

如图 3.1 所示的是 Kotlin 集合和接口的继承关系。

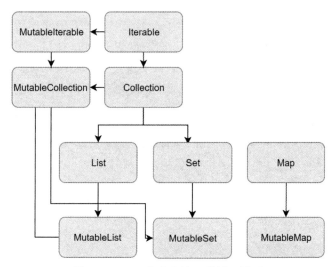

图3.1　Kotlin集合和接口的继承关系

Collection<T>是集合层次结构的根。此接口表示一个只读集合的共同行为：检索大小、检测是否为成员等。Collection 继承自 Iterable<T>接口，它定义了迭代元素的操作。可以使用 Collection 作为适用于不同集合类型的函数的参数。MutableCollection 是一个具有写操作的 Collection 接口，例如 add 及 remove。

List<T>以指定的顺序存储元素，并提供使用索引访问元素的方法。索引从 0 开始，直到最后一个元素的索引，即（list.size－1）。List 中的元素（包括空值）可以重复：List 可以包含任意数量的相同对象或单个对象。如果两个 List 在相同的位置具有相同大小和相同结构的元素，则认为它们是相等的。MutableList 是可以进行写操作的 List，例如，用于在特定位置添加或删除元素。在某些方面，List 与数组（Array）非常相似。但是，它们有一个重要的区别——数组的大小是在初始化时定义的，永远不会改变；而 List 未预定义大小。

作为写操作的结果，可以更改 List 的大小——添加、更新或删除元素。

Set<T>存储唯一的元素；元素的顺序通常未定义。null 元素也是唯一的，一个 Set 中只能包含一个 null。当两个 Set 具有相同的大小并且一个 Set 中的每个元素都能在另一个 Set 中存在相同元素，则两个 Set 相等。MutableSet 是一个带有来自 MutableCollection 的写操作接口的 Set。Set 的默认实现是 LinkedHashSet，保持元素插入时的顺序。另一种实现方式是 HashSet，不声明元素的顺序。

Map<K, V>不是 Collection 接口的继承者，但它也是 Kotlin 的一种集合类型。Map 存储键值对（或条目）；键是唯一的，但是不同的键可以与相同的值配对。Map 接口提供特定的函数进行通过键访问值、搜索键和值等操作。无论键值对的顺序如何，包含相同键值对的两个 Map 是相等的。MutableMap 是一个具有写操作的 Map 接口，可以使用该接口添加一个新的键值对或更新给定键的值。Map 的默认实现是 LinkedHashMap，迭代 Map 时保持元素插入时的顺序。反之，另一种实现方式是 HashMap，不声明元素的顺序。

```
1.  val bob = Person("Bob", 31)
2.  val people = listOf<Person>(Person("Adam", 20), bob, bob)
3.  val people2 = listOf<Person>(Person("Adam", 20), Person("Bob", 31), bob)
4.  // 打印：false
5.  println(people == people2)
6.  bob.age = 32
7.  // 打印：false
8.  println(people == people2)
9.  // 可变 List
10. val numbers = mutableListOf(1, 2, 3, 4)
11. numbers.add(5)
12. numbers.removeAt(1)
13. numbers[0] = 0
14. // 打乱元素顺序
15. numbers.shuffle()
16. // 打印：[4, 5, 0, 3]
17. println(numbers)
18. // Set
19. val numbersBackwards = setOf(4, 3, 2, 1)
20. println("The sets are equal: ${numbers == numbersBackwards}")
21. 
22. // Map
```

```
23. val numbersMap = mapOf("key1" to 1, "key2" to 2, "key3" to 3, "key4" to 1)
24. // 打印: All keys: [key1, key2, key3, key4]
25. println("All keys: ${numbersMap.keys}")
26. // 打印: All values: [1, 2, 3, 1]
27. println("All values: ${numbersMap.values}")
28.
29. val anotherMap = mapOf("key2" to 2, "key1" to 1, "key4" to 1, "key3" to 3)
30. // 打印: The maps are equal: true
31. println("The maps are equal: ${numbersMap == anotherMap}")
32. // 可变 Map
33. val mutableNumbersMap = mutableMapOf("one" to 1, "two" to 2)
34. mutableNumbersMap.put("three", 3)
35. mutableNumbersMap["one"] = 11
```

创建集合的最常用方法是使用标准库函数 listOf<T>()、setOf<T>()、mutableListOf<T>()、mutableSetOf<T>()。如果以逗号分隔的集合元素列表作为参数，编译器会自动检测元素类型。创建空集合时，必须明确指定集合类型。

同样地，Map 也有这样的函数——mapOf() 与 mutableMapOf()。映射的键和值作为 Pair 对象传递（通常使用中缀函数 to 创建）。可以创建可写 Map 并使用写入操作填充它。apply() 函数在初始化时使用。

创建没有任何元素的集合的函数有 emptyList()、emptySet() 与 emptyMap()。创建空集合时，应指定集合将包含的元素类型。对于 List，有一个接收 List 的大小与初始化函数的构造函数，该初始化函数根据索引定义元素的值。

```
1.  // 初始化集合
2.  val numbersSet = setOf("one", "two", "three", "four")
3.  // 初始化空集合
4.  val emptySet = mutableSetOf<String>()
5.  // 初始化 Map
6.  val numbersMap1 = mapOf("key1" to 1, "key2" to 2, "key3" to 3, "key4" to 1)
7.  // 初始化可变 Map
8.  val numbersMap2 = mutableMapOf<String, String>().apply { this["one"] = "1";
    this["two"] = "2" }
9.  // 初始化空 List
10. val empty = emptyList<String>()
11. val doubled = List(3, { it * 2 })
```

```
12. // 打印: [0, 2, 4]
13. println(doubled)
```

要创建与现有集合具有相同元素的集合,可以使用复制操作。标准库中的集合复制操作创建了具有相同元素引用的浅复制集合。因此,对集合元素所做的更改会反映在其所有副本中。但是 toList()、toMutableList()、toSet()等创建了一个具有相同元素的新集合,如果在源集合中添加或删除元素,则不会影响副本。副本也可以独立于源集合进行更改。这些函数还可用于将集合转换为其他类型,例如,根据 List 构建 Set,反之亦然。

```
1. val sourceList = mutableListOf(1, 2, 3)
2. // 复制元素到新 List
3. val copyList = sourceList.toMutableList()
4. sourceList.add(4)
5. // 打印: Copy size: 3
6. println("Copy size: ${copyList.size}")
```

Kotlin 支持使用迭代器遍历集合中的元素,迭代器可以在不暴露集合内部结构的情况下,顺序遍历集合中的元素。迭代器适用于一个一个处理集合元素的场景。Set 和 List 的 iterator()函数提供了迭代器,迭代器初始指向集合的第一个元素,next()函数返回当前元素并指向下一个元素。迭代器可以遍历集合中的元素,但不能获取元素。如果需要再次遍历集合,需要重新创建一个迭代器。此外,用 for 或者 forEach 语句也可以遍历集合中的元素。

```
1. val numberStrs = listOf("one", "two", "three", "four")
2. // 遍历集合元素
3. for (item in numberStrs) {
4.     println(item)
5. }
6. numberStrs.forEach {
7.     println(it)
8. }
```

List 有一个特殊的迭代器 ListIterator,它支持双向遍历,后续遍历用 hasPrevious()、previous(),可以用 nextIndex()和 previousIndex()获取元素下标。

可变集合提供 MutableIterator,当遍历集合时,可以用 remove 移除元素。MutableListIterator 可以修改、添加元素。

```
1. val numberStrs1 = mutableListOf("one", "two", "three", "four")
```

```
2.  val mutableIterator = numberStrs1.iterator()
3.  // 使用迭代器操作集合元素
4.  mutableIterator.next()
5.  mutableIterator.remove()
6.  // 打印: After removal: [two, three, four]
7.  println("After removal: $numberStrs1")
8.  val mutableListIterator = numberStrs1.listIterator()
9.  mutableListIterator.next()
10. mutableListIterator.add("two")
11. mutableListIterator.next()
12. mutableListIterator.set("three")
13. // 打印: [two, two, three, four]
14. println(numberStrs1)
```

Kotlin 可以使用 rangeTo 创建一连串数值，通常用 .. 替代 rangeTo。这些数值可以用 for 语句遍历。如果要进行后续遍历，可以用 downTo。这些数值的间隔通常是 1，也可以用 step 指定步长。如果不需要遍历最后一个元素，可以用 until。可以用 in、!in 判断某个元素是否在区间内。

```
1.  // 打印 1 3 5 7 9, 不包含 10
2.  for (i in 1 until 10 step 2) {
3.      print(i)
4.  }
5.  val rangeInt = 1..10 step 2
6.  // 打印: true
7.  println(3 in rangeInt)
8.  // 打印: true
9.  println(2 !in rangeInt)
10. }
```

Kotlin 提供序列类型 Sequence<T>。当迭代过程包含多个步骤时，每一步执行完，会产生一个结果，下面的步骤会在这个结果的基础上执行。序列中的每一个元素都要依次执行操作步骤，迭代器对集合中的所有元素执行完一个操作过程后再执行下一个。序列可避免产生中间结果，可以提高整个处理链的性能。

通常可以用 sequenceOf() 构建序列；可以用 asSequence() 将 List、Set 转换为序列；可以用 generateSequence() 显示指定序列的第一个元素，函数返回 null 时序列停止增长；序列也可以用组块生成，函数包含 Lambda 表达式，包含 yield()、yieldAll()函数。yield()的参数是一个元素，yieldAll()的参数可以是一个集合，也可以是一个序列，这个序列可以无限长。

```kotlin
1.  // 用函数构建序列
2.  val oddNumbers = generateSequence(1) { it + 2 }
3.  // 打印：[1, 3, 5, 7, 9]
4.  println(oddNumbers.take(5).toList())
5.
6.  // 限长序列
7.  val oddNumbersLessThan10 = generateSequence(1) { if (it < 10) it + 2 else null }
8.  // 打印：6
9.  println(oddNumbersLessThan10.count())
10. val oddNumbers1 = sequence {
11.     yield(1)
12.     yieldAll(listOf(3, 5))
13.     yieldAll(generateSequence(7) { it + 2 })
14. }
15. // 打印：[1, 3, 5, 7, 9]
16. println(oddNumbers1.take(5).toList())
```

序列支持无状态的操作，如 filter、take、map、drop 等。下面的例子展示了序列的处理过程，对每个元素都会执行 filter、map、take 方法，在找出 4 个元素后，不再遍历其余元素。

```kotlin
1.  val words = "The quick brown fox jumps over the lazy dog".split(" ")
2.  // 将集合转换为序列
3.  val wordsSequence = words.asSequence()
4.  val lengthsSequence = wordsSequence
5.         .filter { println("filter: $it"); it.length > 3 }
6.         .map { println("length: ${it.length}"); it.length }
7.         .take(4)
8.  // 打印
9.  // filter: The
10. // filter: quick
11. // length: 5
12. // filter: brown
13. // length: 5
```

```
14. // filter: fox
15. // filter: jumps
16. // length: 5
17. // filter: over
18. // length: 4
19. // [5, 5, 5, 4]
20. println(lengthsSequence.toList())
```

3.4.2 集合操作

Kotlin 标准库提供了很多函数操作集合，如 set、add、search、sorting、filtering、transformations 等。

集合有如下操作：转换（transformations）、过滤（filtering）、加减（plus and minus operators）、分组（grouping）、获取部分集合（retrieving collection parts）、排序（ordering）、聚合操作（aggregate operations）。

集合操作不会改变集合本身，而是生成一个新的集合存放处理后的结果。此外，还可以指定一个可变对象存放处理后的结果。

```
1. val numbers = listOf("one", "two", "three", "four")
2. val filterResults = mutableListOf<String>()
3. // 对 numbers 进行过滤操作，复制到新集合
4. numbers.filterTo(filterResults) { it.length > 3 }
5. // 打印：[three, four]
6. println(filterResults)
```

Kotlin 提供了很多集合转换的扩展操作，这些操作会产生新的集合。下面介绍几种转换操作：映射（mapping），用 Lambda 表达式对当前集合中的元素进行处理，生成一个新的集合，两个集合中元素的顺序一致。当产生 null 元素时，可以用 mapNotNull() 函数将其过滤掉。当转换映射时，可以用 mapKeys 处理 key 值，或者用 mapValues 处理 value 值。

```
1. val numbers1 = setOf(1, 2, 3)
2. // 集合映射操作
3. println(numbers1.map { it * 3 })
4. println(numbers1.mapIndexed { idx, value -> value * idx })
5. // 用 mapNotNull 函数过滤 null 元素
```

```kotlin
6.  println(numbers1.mapNotNull { if ( it == 2) null else it * 3 })
7.  println(numbers1.mapIndexedNotNull { idx, value -> if (idx == 0) null else value
    * idx })
8.
9.  val numbersMap = mapOf("key1" to 1, "key2" to 2, "key3" to 3, "key11" to 11)
10. // 打印：{KEY1=1, KEY2=2, KEY3=3, KEY11=11}
11. println(numbersMap.mapKeys { it.key.toUpperCase() })
12. // 打印：{key1=5, key2=6, key3=7, key11=16}
13. println(numbersMap.mapValues { it.value + it.key.length })
```

双路合并（zip）操作，将两个集合中的元素合并为一个 List，List 中每个元素成对出现，是 Pair 类型的。如果两个集合大小不一样，zip 操作取较小的集合长度。此外，在 zip 操作的基础上，还可以进行 Lambda 操作，最终返回的 List 元素集合就是由 Lambda 操作后的元素组成的。相反，使用 unzip 函数可以进行反向操作。

```kotlin
1.  val colors = listOf("red", "brown", "grey")
2.  val animals = listOf("fox", "bear", "wolf")
3.  // 使用 zip 合并集合
4.  // 打印：[(red, fox), (brown, bear), (grey, wolf)]
5.  println(colors zip animals)
6.
7.  val twoAnimals = listOf("fox", "bear")
8.  // 打印：[(red, fox), (brown, bear)]
9.  println(colors.zip(twoAnimals))
10. // 打印：[The Fox is red, The Bear is brown, The Wolf is grey]
11. println(colors.zip(animals) { color, animal -> "The ${animal.capitalize()} is
    $color"})
12. val numberPairs = listOf("one" to 1, "two" to 2, "three" to 3, "four" to 4)
13. // 打印：([one, two, three, four], [1, 2, 3, 4])
14. println(numberPairs.unzip())
```

关联（association）操作，可以对集合中的元素进行处理，以生成一个映射。基本的关联函数是 associateWith()，集合中的元素作为映射的 key，转换后的元素作为映射的 value。associateBy() 函数将集合中的元素作为映射的 value，转换后的元素作为映射的 key。

```kotlin
1.  val numbers2 = listOf("one", "two", "two", "four")
2.  // 集合关联，打印：{one=3, two=3, four=4}
```

```
3.    println(numbers2.associateWith { it.length })
4.    // 打印：{O=one, T=two, F=four}
5.    println(numbers2.associateBy { it.first().toUpperCase() })
6.    // 打印：{O=3, T=3, F=4}
7.    println(numbers2.associateBy(keySelector = { it.first().toUpperCase() },
      valueTransform = { it.length }))
```

打平（flattern）操作，flattern()函数将若干集合中的元素放在一个集合中；flatternMap()函数对若干集合进行映射操作，然后再打平。

```
1.    val numberSets = listOf(setOf(1, 2, 3), setOf(4, 5, 6), setOf(1, 2))
2.    // 使用 flattern 打平集合，打印：[1, 2, 3, 4, 5, 6, 1, 2]
3.    println(numberSets.flatten())
```

字符串处理操作，可以用 joinToString()函数将集合中的元素生成一个字符串，用 joinTo()函数将生成的字符串拼接在另一个字符串后。

```
1.    val numbers3 = listOf("one", "two", "three", "four")
2.    // 打印：one, two, three, four
3.    println(numbers3.joinToString())
4.    val listString = StringBuffer("The list of numbers: ")
5.    numbers3.joinTo(listString)
6.    // 打印：The list of numbers: one, two, three, four
7.    println(listString)
```

过滤操作在集合处理中使用广泛。Kotlin 使用谓词函数实现过滤操作，使用 Lambda 表达式处理集合的元素，并返回一个布尔值。如果需要使用集合元素的下标，可以使用 filterIndexed()函数。如果需要使用相反条件过滤，可以使用 filterNot()函数。如果需要过滤类型，可以使用 filterIsInstance<T>()函数。需要过滤空，可以使用 filterNotNull()函数。

```
1.    val numbersMap1 = mapOf("key1" to 1, "key2" to 2, "key3" to 3, "key11" to 11)
2.    val filteredMap = numbersMap1.filter { (key, value) -> key.endsWith("1") && value > 10}
3.    // 打印：{key11=11}
4.    println(filteredMap)
5.    val numbers4 = listOf("one", "two", "three", "four")
6.    val filteredIdx = numbers4.filterIndexed { index, s -> (index != 0) && (s.length
      < 5) }
7.    val filteredNot = numbers4.filterNot { it.length <= 3 }
```

```
8.  // 打印：[two, four]
9.  println(filteredIdx)
10. // 打印：[three, four]
11. println(filteredNot)
```

可以使用 partition() 函数对集合进行划分，一部分满足过滤条件，另一部分不满足过滤条件。可以使用 any()（至少有一个）、none()（一个也没有）、all()（全部满足）来检验谓词函数。

集合还可以进行加减操作，运算符的左值是集合，右值可以是一个元素或者集合。

```
1. val numbers5 = listOf("one", "two", "three", "four")
2.
3. val plusList = numbers5 + "five"
4. val minusList = numbers5 - listOf("three", "four")
5. // 打印：[one, two, three, four, five]
6. println(plusList)
7. // 打印：[one, two]
8. println(minusList)
```

Kotlin 还提供了分组操作。groupBy 函数使用一个 Lambda 表达式，返回一个映射。映射的 key 是 Lambda 表达式返回的值，value 是集合中的元素。

```
1. val numbers6 = listOf("one", "two", "three", "four", "five")
2. // 打印：{O=[one], T=[two, three], F=[four, five]}
3. println(numbers6.groupBy { it.first().toUpperCase() })
4. // 打印：{o=[ONE], t=[TWO, THREE], f=[FOUR, FIVE]}
5. println(numbers6.groupBy(keySelector = { it.first() }, valueTransform =
   { it.toUpperCase() }))
```

Kotlin 提供了丰富的操作以获取集合的一部分。slice() 函数根据给定下标返回集合中的元素。take() 函数从第一个元素开始，截取给定个数的集合元素；如果截取个数大于集合长度，返回整个集合元素。drop() 函数从第一个元素开始，丢弃给定个数的元素，然后返回其余元素。如果要使用谓词函数，可以用 takeWhile()、takeLastWhile()、dropWhile() 函数、dropLastWhile()。chunked() 函数将集合分块，遍历集合元素，达到给定数量，生成一个 List，直到最后一个元素。windowed() 函数从第一个元素开始进行遍历，可以指定窗口大小。此外，windowed() 函数可以指定步长 step 等参数。如果滑动窗口中只有两个元素，可以用 zipWithNext() 函数。

```kotlin
1.  val numbers7 = listOf("one", "two", "three", "four", "five", "six")
2.  // 打印: [two, three, four]
3.  println(numbers7.slice(1..3))
4.  // 打印: [one, three, five]
5.  println(numbers7.slice(0..4 step 2))
6.  // 打印: [four, six, one]
7.  println(numbers7.slice(setOf(3, 5, 0)))
8.  val numbers8 = listOf("one", "two", "three", "four", "five", "six")
9.  // 打印: [one, two, three]
10. println(numbers8.take(3))
11. // 打印: [four, five, six]
12. println(numbers8.takeLast(3))
13. // 打印: [two, three, four, five, six]
14. println(numbers8.drop(1))
15. // 打印: [one]
16. println(numbers8.dropLast(5))
17. // 打印: [one, two, three]
18. println(numbers8.takeWhile { !it.startsWith('f') })
19. // 打印: [four, five, six]
20. println(numbers8.takeLastWhile { it != "three" })
21. // 打印: [three, four, five, six]
22. println(numbers8.dropWhile { it.length == 3 })
23. // 打印: [one, two, three, four]
24. println(numbers8.dropLastWhile { it.contains('i') })
25. val numbers9 = (0..13).toList()
26. // 打印: [[0, 1, 2], [3, 4, 5], [6, 7, 8], [9, 10, 11], [12, 13]]
27. println(numbers9.chunked(3))
28. // 打印: [3, 12, 21, 30, 25]
29. println(numbers9.chunked(3) { it.sum() })
30. val numbers10 = (1..10).toList()
31. // 打印: [[1, 2, 3], [2, 3, 4], [3, 4, 5], [4, 5, 6], [5, 6, 7], [6, 7, 8], //[7, 8, 9], [8, 9, 10]]
32. println(numbers10.windowed(3))
33. // 打印: [[1, 2, 3], [3, 4, 5], [5, 6, 7], [7, 8, 9], [9, 10]]
34. println(numbers10.windowed(3, step = 2, partialWindows = true))
35. // 打印: [6, 9, 12, 15, 18, 21, 24, 27]
```

```kotlin
36. println(numbers10.windowed(3) { it.sum() })
37. // 打印：[(1, 2), (2, 3), (3, 4), (4, 5), (5, 6), (6, 7), (7, 8), (8, 9), (9, //10)]
38. println(numbers10.zipWithNext())
39. // 打印：[3, 5, 7, 9, 11, 13, 15, 17, 19]
40. println(numbers10.zipWithNext() { s1, s2 -> s1 + s2})
```

如果要获取集合中的单个元素，可以使用 elementAt() 按位置获取，因 first() 获取第一个元素，用 last() 获取最后一个元素，也可以在 first() 或者 last() 后用谓词函数按条件获取。用 random() 函数可随机获取元素，用 contains() 检测是否存在某个元素，用 containsAll() 检测是否存在多个元素，用 isEmpty()、isNotEmpty() 判断集合是否为空。

Kotlin 可以用 Comparable 接口对自定义的类型进行排序，Kotlin 自带的类型默认支持排序，数值型按数值大小排序，字符型按照字母顺序排序。此外，还可以用 Comparator 自定义顺序，compareBy() 是 Comparator 的简单写法。自然顺序可以用 sorted() 和 sortedDescending() 进行升序或降序操作。自定义顺序可以用 sortedBy() 和 sortedByDescending() 进行升序或降序操作。倒序可以用 reversed()、asReversed()。随机顺序可以用 shuffled()。

```kotlin
1.  val unSortednumbers = listOf("one", "two", "three", "four")
2.
3.  val lengthComparator = Comparator { str1: String, str2: String -> str1.length - str2.length }
4.  // 打印: [one, two, four, three]
5.  println(unSortednumbers.sortedWith(lengthComparator))
6.  // 打印: [one, two, four, three]
7.  println(unSortednumbers.sortedWith(compareBy { it.length }))
8.  // 打印: Sorted ascending: [four, one, three, two]
9.  println("Sorted ascending: ${unSortednumbers.sorted()}")
10. // 打印: Sorted descending: [two, three, one, four]
11. println("Sorted descending: ${unSortednumbers.sortedDescending()}")
12.
13. val sortedNumbers = unSortednumbers.sortedBy { it.length }
14. // 打印: Sorted by length ascending: [one, two, four, three]
15. println("Sorted by length ascending: $sortedNumbers")
16. val sortedByLast = unSortednumbers.sortedByDescending { it.last() }
17. // 打印: Sorted by the last letter descending: [four, two, one, three]
18. println("Sorted by the last letter descending: $sortedByLast")
19. // 打印：[four, three, two, one]
```

```
20. println(unSortednumbers.reversed())
21. // 打印：[two, three, four, one]
22. println(unSortednumbers.shuffled())
```

集合的聚合操作函数有求最小值 min()、最大值 max()、平均值 average()、求和 sum()、计数 count()，带有函数的求最大值/最小值为 maxBy()/minBy()，Comparator 对象的求最大值/最小值为 maxWith()/minWith()，带有谓词函数的求和为 sumBy()，返回 Double 类型的求和为 sumDouble()，累加为 fold()、reduce()，fold()有初始值，reduce()开始时用前两个元素作为参数。

```
1.  val numbersMerge = listOf(6, 42, 10, 4)
2.  // 打印：Count: 4
3.  println("Count: ${numbersMerge.count()}")
4.  // 打印：Max: 42
5.  println("Max: ${numbersMerge.max()}")
6.  // 打印：Min: 4
7.  println("Min: ${numbersMerge.min()}")
8.  // 打印：Average: 15.5
9.  println("Average: ${numbersMerge.average()}")
10. // 打印：Sum: 62
11. println("Sum: ${numbersMerge.sum()}")
12. val min3Remainder = numbersMerge.minBy { it % 3 }
13. // 打印：6
14. println(min3Remainder)
15. val max3Remainder = numbersMerge.maxWith(compareBy { it - 10 })
16. // 打印：42
17. println(max3Remainder)
18. // 打印：124
19. println(numbersMerge.sumBy { it * 2 })
20. // 打印：31.0
21. println(numbersMerge.sumByDouble { it.toDouble() / 2 })
22. val sum = numbersMerge.reduce { sum, element -> sum + element }
23. // 打印：62
24. println(sum)
25. val sumDoubled = numbersMerge.fold(0) { sum, element -> sum + element * 2 } // 打印：124
26. println(sumDoubled)
```

集合可以通过 add() 添加单个元素，addAll() 添加多个元素，remove() 删除元素，retainAll() 保留符合条件的元素，clear() 清空集合，minusAsign() 和 minus() 删除元素。

3.4.3　List、Set、Map 相关操作

List 是使用广泛的集合，List 可以按索引取元素，getOrElse() 函数若取不到元素就返回默认值，getOrNull() 函数若取不到元素就返回 null。对于有序 List，可以用 binarySearch() 进行二分查找，此外，可以自定义排序规则。

```
1.  val listNuumbers = listOf(1, 2, 3, 2)
2.  // 打印: 1
3.  println(listNuumbers.get(0))
4.  // 打印: 1
5.  println(listNuumbers[0])
6.  // 打印: null
7.  println(listNuumbers.getOrNull(5))            // null
8.  // 打印: 5
9.  println(listNuumbers.getOrElse(5, {it}))      // 5
10. // 打印: [1, 2]
11. println(listNuumbers.subList(0, 2))
12. // 打印: 1
13. println(listNuumbers.indexOf(2))
14. // 打印: 3
15. println(listNuumbers.lastIndexOf(2))
16. // 打印: 2
17. println(listNuumbers.indexOfFirst { it > 2})
18. // 打印: 2
19. println(listNuumbers.indexOfLast { it % 2 == 1})
20.
21. listNuumbers.sorted()
22. // 打印: 1
23. println(listNuumbers.binarySearch(2))   // 3
24. // 打印: -5
25. println(listNuumbers.binarySearch(4))  // -5
26. // 打印: 0
27. println(listNuumbers.binarySearch(1, 0, 2))  // -3
```

Set 提供了求交集 intersect()、合并 union()、剔除交集元素 subtract()等操作。

```kotlin
1.  val setNnumbers = setOf("one", "two", "three")
2.  // 打印：[one, two, three, four, five]
3.  println(setNnumbers union setOf("four", "five"))
4.  // 打印：[one, two]
5.  println(setNnumbers intersect setOf("two", "one"))
6.  // 打印：[one, two]
7.  println(setNnumbers subtract setOf("three", "four"))
8.  // 打印：[one, two]
9.  println(setNnumbers subtract setOf("four", "three")) // same output
```

Map 提供了取键、值操作，过滤键、值操作，plus 和 minus 操作，以及添加和更新操作。

```kotlin
1.  val mapNumbers = mapOf("one" to 1, "two" to 2, "three" to 3)
2.  // 打印：1
3.  println(mapNumbers.get("one"))
4.  // 打印：1
5.  println(mapNumbers["one"])
6.  // 打印：10
7.  println(mapNumbers.getOrDefault("four", 10))
8.  // 打印：null
9.  println(mapNumbers["five"])                  // null
10. // 打印：[one, two, three]
11. println(mapNumbers.keys)
12. // 打印：[1, 2, 3]
13. println(mapNumbers.values)
14. // 打印：{}
15. println(mapNumbers.filter { (key, value) -> key.endsWith("1") && value > 10})
16. // 打印：{}
17. println(mapNumbers.filterKeys { it.endsWith("1") })
18. // 打印：{one=1, two=2, three=3}
19. println(mapNumbers.filterValues { it < 10 })
20. // 打印：{one=1, two=2, three=3, four=4}
21. println(mapNumbers + Pair("four", 4))
22. // 打印：{one=10, two=2, three=3}
23. println(mapNumbers + Pair("one", 10))
24. // 打印：{one=11, two=2, three=3, five=5}
```

```
25. println(mapNumbers + mapOf("five" to 5, "one" to 11))
26. // 打印：{two=2, three=3}
27. println(mapNumbers - "one")
28. // 打印：{one=1, three=3}
29. println(mapNumbers - listOf("two", "four"))
```

3.5 协程

本节介绍 Kotlin 中协程的基础概念和进阶用法。

3.5.1 协程基础

kotlinx.coroutines 是 Kotlin 的协程库。下面是一个协程的例子：

```
1.  import kotlinx.coroutines.*
2.  fun main() {
3.      // 在后台启动一个新的协程并继续
4.      GlobalScope.launch {
5.          // 非阻塞地等待 1 秒（默认时间单位是毫秒）
6.          delay(1000L)
7.          // 在延迟后打印输出
8.          println("World!")
9.      }
10.     // 协程已在等待时，主线程还在继续
11.     println("Hello,")
12.     // 但是这个表达式阻塞了主线程
13.     runBlocking {
14.         // ……延迟 2 秒来保证 JVM 的存活
15.         delay(2000L)
16.     }
17. }
```

本质上，协程是轻量级的线程。它们在某些 coroutineScope 上下文中与 launch 协程构建器一起启动。GlobalScope 像守护线程，在 GlobalScope 中启动的活动协程并不会使进程保活。可以将 GlobalScope.launch{…}替换为 thread{…}，将 delay(…)替换为 Thread.sleep(…)来达到同样的目的。delay 是一个特殊的挂起函数，它不会造成线程阻塞，但是会挂起协程，

并且只能在协程中使用。可显式地用 runBlocking 协程构建器来阻塞线程。可以用 join 显式（以非阻塞方式）地等待所启动的后台作业执行结束。可以在执行操作所在的指定作用域内启动协程，而不是像通常使用线程（线程总是全局的）那样在 GlobalScope 中启动。还可以使用 coroutineScope 构建器声明自己的作用域。

runBlocking 与 coroutineScope 的主要区别在于，后者在等待所有子协程执行完毕时不会阻塞当前线程。如果将 launch{…}内部的代码块提取到独立的函数中，需要用 suspend 修饰。

取消协程的执行

协程是可以被取消的，可以用 cancel()取消协程。协程的取消是协作的。一段协程代码必须协作才能被取消。运行下面的示例代码，可以看到它连续打印出了"I'm sleeping"，甚至在调用取消后，作业仍然执行了 5 次循环迭代并运行到结束为止。

```
1.  fun main() {
2.      runBlocking {
3.          val startTime = System.currentTimeMillis()
4.          val job = launch(Dispatchers.Default) {
5.              var nextPrintTime = startTime
6.              var i = 0
7.              while (i < 5) {
8.                  // 一个执行计算任务的循环，只是为了占用 CPU
9.                  // 每秒打印消息两次
10.                 if (System.currentTimeMillis() >= nextPrintTime) {
11.                     println("job: I'm sleeping ${i++} ...")
12.                     nextPrintTime += 500L
13.                 }
14.             }
15.         }
16.         // 等待一段时间
17.         delay(1300L)
18.         println("main: I'm tired of waiting!")
19.         // 取消一个作业并且等待它结束
20.         job.cancelAndJoin()
21.         println("main: Now I can quit.")
22.     }
23. }
```

计算任务是可以取消的，可以显式检查取消状态。将 while(i<5)替换为 while(isActive)，在循环到第三次后，任务就被取消了。当协程取消时，可以在 finally()函数中释放资源。取消一个协程的理由是它可能超时，可以使用 withTimeout()、withTimeoutOrNull()函数在协程超时后取消协程。

```
1.  // 打印: I'm sleeping 0 ...
2.  // I'm sleeping 1 ...
3.  // I'm sleeping 2 ...
4.  withTimeout(1300L) {
5.      repeat(1000) { i ->
6.             println("I'm sleeping $i ...")
7.             delay(500L)
8.      }
9.  }
```

组合挂起函数

如果要根据第一个函数的结果来决定是否需要调用第二个函数或者决定如何调用第二个函数，可使用普通的顺序进行调用。下面列举的例子的耗时是两个函数耗时之和。

```
1.  suspend fun doSomethingUsefulOne(): Int {
2.      delay(1000L) // 假设我们在这里做了一些有用的事
3.      return 13
4.  }
5.
6.  suspend fun doSomethingUsefulTwo(): Int {
7.      delay(1000L) // 假设我们在这里也做了一些有用的事
8.      return 29
9.  }
10.
11. fun main() = runBlocking<Unit> {
12.     val time = measureTimeMillis {
13.         val one = doSomethingUsefulOne()
14.         val two = doSomethingUsefulTwo()
15.         // 打印: The answer is 42
16.         println("The answer is ${one + two}")
17.     }
```

```
18.     // 打印 Completed in 2010 ms
19.     println("Completed in $time ms")
20. }
```

当然如果两个函数没有依赖,可以用 async 进行并发。async 返回一个 Deferred,一个轻量级的非阻塞 future,可以使用.await()(经过一定的时间)得到它的最终结果,但是 Deferred 也是一个作业,是可以取消的。

```
1. val time = measureTimeMillis {
2.     val one = async { doSomethingUsefulOne() }
3.     val two = async { doSomethingUsefulTwo() }
4.     // 打印: The answer is 42
5.     println("The answer is ${one.await() + two.await()}")
6. }
7. // 打印: Completed in 1027 ms
8. println("Completed in $time ms")
```

可以使用 async 进行结构化并发,如下例所示。如果在 concurrentSum()函数内部发生了错误,并且抛出了一个异常,那么所有在作用域中启动的协程都会被取消。取消始终通过协程的层次结构来进行传递。

```
1. suspend fun concurrentSum(): Int = coroutineScope {
2.     val one = async { doSomethingUsefulOne() }
3.     val two = async { doSomethingUsefulTwo() }
4.     one.await() + two.await()
5. }
```

协程上下文与调度器

协程总是运行在一些以 CoroutineContext 类型为代表的上下文中,它们被定义在 Kotlin 的标准库里。协程上下文包含一个协程调度器,它确定了哪些线程或与线程相对应的协程执行。协程调度器可以将协程限制在一个特定的线程内执行,或将其分派到一个线程池,或让其不受限地运行。

```
1. fun main() = runBlocking<Unit> {
2.     launch {
3.         // 运行在父协程的上下文中,即 runBlocking 主协程
4.         println("main runBlocking      : I'm working in thread
```

```
5.                ${Thread.currentThread().name}")
6.            }
7.            launch(Dispatchers.Unconfined) {
8.                // 不受限的——将工作在主线程中
9.                println("Unconfined            : I'm working in thread
              ${Thread.currentThread().name}")
10.           }
11.           launch(Dispatchers.Default) {
12.               // 将会获取默认调度器
13.               println("Default               : I'm working in thread
              ${Thread.currentThread().name}")
14.           }
15.           launch(newSingleThreadContext("MyOwnThread")) {
16.               // 将获得一个新的线程
17.               println("newSingleThreadContext: I'm working in thread
              ${Thread.currentThread().name}")
18.           }
19.       }
```

当调用 launch {…} 时不传参数，它会从启动了它的 CoroutineScope 中继承上下文（及调度器）。在上述例子中，它从 main 线程中的 runBlocking 主协程继承了上下文。

Dispatchers.Unconfined 是一个特殊的调度器且似乎也运行在 main 线程中，但实际上，它是一种不同的机制。

当协程在 GlobalScope 中启动的时候使用，代表 Dispatchers.Default 使用了共享的后台线程池，所以 GlobalScope.launch {…} 也可以使用相同的调度器 launch(Dispatchers. Default) {…}。

newSingleThreadContext 为协程的运行启动了一个线程。专用线程是一种非常昂贵的资源。在真实的应用程序中协程和线程都必须被释放，当不再需要的时候，使用 close()函数释放，或将 newSingleThreadContext 存储在一个顶层变量中使其在整个应用程序中被重用。

3.5.2 协程进阶

Kotlin 的流（Flow）可以处理挂起函数异步返回的多个值。如下面的示例就定义了一个流，这个流不阻塞主线程，用 emit()函数发射流，用 collect()函数收集流。flow{}构建块的代码可以挂起，函数 foo()不再有 suspend 标识符。流采用与协程同样的协作取消方式——withTimeoutOrNull，当超时时取消并停止执行其代码。

```
1.  fun foo(): Flow<Int> = flow {
2.      // 流构建器
3.      for (i in 1..3) {
4.          delay(100) // 假设我们在这里做了一些有用的事情
5.          emit(i) // 发送下一个值
6.      }
7.  }
8.  fun main() = runBlocking<Unit> {
9.      // 启动并发的协程以验证主线程并未阻塞
10.     launch {
11.         for (k in 1..3) {
12.             println("I'm not blocked $k")
13.             delay(100)
14.         }
15.     }
16.     // 收集这个流
17.     foo().collect { value -> println(value) }
18. }
```

可以用 flowOf()定义一个发射固定值集的流，用.asFlow()将各种集合与序列转换为流。流的过渡操作符，如 map、filter，应用于上游流，返回下游流。流的转换操作符，常用的如 transform，可以实施更复杂的转换。限长过渡操作符，如 take，在流触及相应限制的时候会将其取消。

```
1.  suspend fun performRequest(request: Int): String {
2.      delay(1000) // 模仿长时间运行的异步工作
3.      return "response $request"
4.  }
5.  fun main() = runBlocking<Unit> {
6.      (1..3).asFlow() // 一个请求流
7.          .map { request -> performRequest(request) }
8.          .collect { response -> println(response) }
9.  }
```

流的末端操作符有 collect、toList 和 toSet，它们可将流转换为集合，first()获取第一个值，single()确保流发射单个值，reduce()、fold()将流规约到单个值。

可以在流上使用 buffer 操作符来并发运行发射元素的代码及收集的代码，而不是顺序运行它们。

```kotlin
1.  fun log(msg: String) = println("[${Thread.currentThread().name}] $msg")
2.  fun foo(): Flow<Int> = flow {
3.      for (i in 1..3) {
4.          Thread.sleep(100) // 假设我们以消耗 CPU 的方式进行计算
5.          log("Emitting $i")
6.          emit(i) // 发射下一个值
7.      }
8.  }.flowOn(Dispatchers.Default) // 在流构建器中改变消耗 CPU 代码上下文的正确方式
9.  fun main() = runBlocking<Unit> {
10.     val time = measureTimeMillis {
11.         foo()
12.             .buffer() // 缓冲发射项，无须等待
13.             .collect { value ->
14.                 delay(300) // 假设我们花费 300 毫秒来处理它
15.                 println(value)
16.             }
17.     }
18.     println("Collected in $time ms")
19. }
```

通道提供了一种在流中传输值的方法。Channel 是和 BlockingQueue 非常相似的概念，不同之处是，它代替了阻塞的 put 操作并提供了挂起的 send，还替代了阻塞的 take 操作并提供了挂起的 receive。可以通过关闭通道来表明没有更多的元素将会进入通道。在接收者中可以定期使用 for 循环来从通道中接收元素。

```kotlin
1.  fun main() = runBlocking<Unit> {
2.      val channel = Channel<Int>()
3.      launch {
4.          // 这里可能是消耗大量 CPU 运算的异步逻辑，我们将仅做 5 次整数的平方运算并发送
5.          for (x in 1..5) channel.send(x * x)
6.          channel.close() // 结束发送
7.      }
```

```
8.      // 这里打印了 5 次接收到的整数
9.      repeat(5) { println(channel.receive()) }
10.     // 这里使用 for 循环来打印所有接收到的元素（直到通道关闭）
11.     for (y in channel) println(y)
12.     println("Done!")
13. }
```

除了 Channel()工厂函数外，还可以用 produce 构造通道。在构造通道时，一个可选的参数是 capacity，它用来指定缓冲区大小。缓冲操作允许发送者在被挂起前发送多个元素，当缓冲区被占满的时候将会引起阻塞。通道发送和接收操作遵守先进先出原则。

```
1. fun CoroutineScope.produceSquares(): ReceiveChannel<Int> = produce {
2.     for (x in 1..5) send(x * x)
3. }
4. fun main() = runBlocking {
5.     val squares = produceSquares()
6.     // 打印: 1 4 9 16 25
7.     squares.consumeEach { println(it) }
8.     println("Done!")
9. }
```

3.6 小结

本章介绍了 Kotlin 的语法，涵盖了常用的基础语法和高阶特性。首先介绍了 Kotlin 的基本类型、包、引用、流程控制、跳转等；接着介绍了 Kotlin 的类、属性、构造函数、接口、泛型；然后介绍了 Kotlin 的函数、Lambda 表达式、集合等，这部分是经常会用到的；最后介绍了协程，从基础的例子开始逐步深入介绍协程，协程更加轻量级，功能也很强大。这些语法是开发微服务应用的基础。下一章我们将介绍如何用 Kotlin 集成各种基础中间件。

第 4 章
Kotlin 在常用中间件中的应用

本章主要介绍 Kotlin 在常用中间件中的应用，通过示例程序，将展示 Kotlin 集成 Spring Boot、Redis、JPA、QueryDSL、MongoDB、Spring Security、RocketMQ、Elasticsearch、Swagger 的方法。读者可以掌握使用 Kotlin 操作常用中间件的技巧。

4.1　Kotlin 集成 Spring Boot

Spring Boot 是由 Pivotal 团队开发的，设计的目的是简化 Spring 应用的初始搭建和开发过程。本节介绍 Kotlin 集成 Spring Boot 开发。

4.1.1　Spring Boot 介绍

从 2014 年 4 月发布 1.0.0.RELEASE 到现在的最新版本 2.2.2.RELEASE，从最初的基于 Spring 4 到现在基于 Spring 5，从同步阻塞编程到异步响应式编程，Spring Boot 经历了数十个 RELEASE 版本，发展迅速，表现稳定，其各版本发行时间如表 4.1 所示。越来越多的企业在生产中使用 Spring Boot 进行企业级应用开发。

表 4.1　Spring Boot、Spring 版本的发行时间

时间	Spring Boot 版本	Spring 版本
2014 年	1.0.x	4.0.x.RELEASE
2014—2015 年	1.1.x	4.0.x.RELEASE
2015 年	1.2.x	4.1.x.RELEASE
2015—2016 年	1.3.x	4.2.x.RELEASE
2016—2017 年	1.4.x	4.3.x.RELEASE
2017—2018 年	1.5.x	4.3.x.RELEASE
2018—2019 年	2.0.x	5.0.x.RELEASE
2018—2020 年	2.1.x	5.1.x.RELEASE
2019—2020 年	2.2.x	5.2.x.RELEASE

Spring Boot 基于约定优于配置的思想，让开发人员不必在配置与逻辑业务之间进行思维的切换。Spring Boot 简化了 Spring 应用的开发，不再需要 XML 配置文件，使用注解方式提高了开发效率。Spring Boot 默认配置了很多框架的使用方式，提供 starter 包，简化配置，开箱即用。Spring Boot 尽可能地根据项目依赖来自动配置 Spring 框架。Spring Boot 提供了可以直接在生产环境中使用的功能，如性能指标、应用信息和应用健康检查。

Spring Boot 内嵌 Tomcat、Jetty、Undertow 等容器，直接用 Jar 包的方式进行部署，而传统的 Spring 应用需要用 war 包方式进行部署。Spring Boot 的部署方法非常简单，一行命令就可以部署一个 Spring Boot 应用；可以很方便地用 Docker、Kubernetes 进行部署，适用于云原生应用，使系统的扩容、运维更加方便。

Spring Boot 广泛应用于企业级应用和微服务开发。Spring Cloud 微服务框架就是在 Spring Boot 基础上开发的。此外，很多开源项目提供了 Spring Boot 的集成，如 rocketmq-spring-boot-starter，方便用户使用。

4.1.2　用 Kotlin 开发一个 Spring Boot 项目

在 Spring 网站上创建一个基于 Maven 的 Kotlin Spring Boot 项目。填写 Group、Artifact，选择依赖的包 Spring Web，然后下载到本地，如图 4.1 所示。

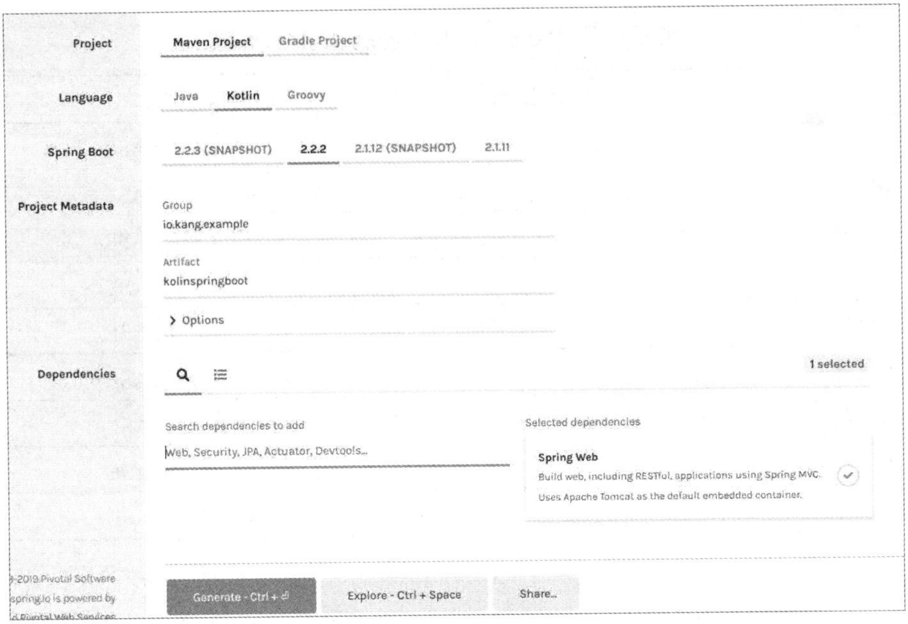

图4.1　Spring Initializr

解压文件，用 IDEA 打开这个工程，可以看到 pom 文件如下：该 pom 文件定义了父依赖，通过父依赖可以自动找到 dependencies 中依赖包的版本号；此外，还指定了 Kotlin 的版本是 1.3.61，Spring Boot 的版本是 2.2.2.RELEASE。

```xml
1.  <?xml version="1.0" encoding="UTF-8"?>
2.  <project xmlns="http://maven.apache.org/POM/4.0.0" xmlns:xsi="http://www.w3.org/2001/XMLSchema-instance"
3.      xsi:schemaLocation="http://maven.apache.org/POM/4.0.0 https://maven.apache.org/xsd/maven-4.0.0.xsd">
4.      <modelVersion>4.0.0</modelVersion>
5.      <!-- 父 pom，定义包的依赖 -->
6.      <parent>
7.          <groupId>org.springframework.boot</groupId>
8.          <artifactId>spring-boot-starter-parent</artifactId>
9.          <version>2.2.2.RELEASE</version>
10.         <relativePath/> <!-- lookup parent from repository -->
11.     </parent>
12.     <!-- 子工程相关信息 -->
13.     <groupId>io.kang.example</groupId>
```

```xml
14.    <artifactId>kolinspringboot</artifactId>
15.    <version>0.0.1-SNAPSHOT</version>
16.    <name>kolinspringboot</name>
17.    <description>Demo project for Spring Boot</description>
18.    <!-- 定义属性 -->
19.    <properties>
20.        <java.version>1.8</java.version>
21.        <kotlin.version>1.3.61</kotlin.version>
22.    </properties>
23.    <dependencies>
24.        <!-- Spring Boot 启动包 -->
25.        <dependency>
26.            <groupId>org.springframework.boot</groupId>
27.            <artifactId>spring-boot-starter</artifactId>
28.        </dependency>
29.        <!-- Kotlin 相关依赖包 -->
30.        <dependency>
31.            <groupId>org.jetbrains.kotlin</groupId>
32.            <artifactId>kotlin-reflect</artifactId>
33.        </dependency>
34.        <dependency>
35.            <groupId>org.jetbrains.kotlin</groupId>
36.            <artifactId>kotlin-stdlib-jdk8</artifactId>
37.        </dependency>
38.        <dependency>
39.            <groupId>org.springframework.boot</groupId>
40.            <artifactId>spring-boot-starter-test</artifactId>
41.            <scope>test</scope>
42.            <exclusions>
43.                <exclusion>
44.                    <groupId>org.junit.vintage</groupId>
45.                    <artifactId>junit-vintage-engine</artifactId>
46.                </exclusion>
47.            </exclusions>
48.        </dependency>
49.    </dependencies>
50.    <build>
```

```xml
51.         <!-- Kotlin 源码路径 -->
52.         <sourceDirectory>${project.basedir}/src/main/kotlin</sourceDirectory>
53.         <testSourceDirectory>${project.basedir}/src/test/kotlin</testSourceDirectory>
54.         <plugins>
55.             <!-- Spring Boot Maven 打包插件 -->
56.             <plugin>
57.                 <groupId>org.springframework.boot</groupId>
58.                 <artifactId>spring-boot-maven-plugin</artifactId>
59.             </plugin>
60.             <!-- Kotlin Maven 插件 -->
61.             <plugin>
62.                 <groupId>org.jetbrains.kotlin</groupId>
63.                 <artifactId>kotlin-maven-plugin</artifactId>
64.                 <configuration>
65.                     <args>
66.                         <arg>-Xjsr305=strict</arg>
67.                     </args>
68.                     <compilerPlugins>
69.                         <plugin>spring</plugin>
70.                     </compilerPlugins>
71.                 </configuration>
72.                 <dependencies>
73.                     <dependency>
74.                         <groupId>org.jetbrains.kotlin</groupId>
75.                         <artifactId>kotlin-maven-allopen</artifactId>
76.                         <version>${kotlin.version}</version>
77.                     </dependency>
78.                 </dependencies>
79.             </plugin>
80.         </plugins>
81.     </build>
82.
83. </project>
```

下面用 Kotlin 编写一个简单的 Spring Boot Web 应用：定义一个 Spring Boot 启动类，加上@SpringBootApplication 注解；定义一个接口，通过 http://localhost:8080/index 可以访问这个接口；相关的配置放在 application.yml 中。

和用 Java 开发 Spring Boot 项目类似，Kotlin 在 main 函数中启动应用，用 GetMapping 定义一个 get 接口，使用 @RestController 后就不用为每个方法添加 @ResponseBody 注解了。Kotlin 的语法更加简洁。

KotlinSpringbootApplication.kt 的代码如下：

```kotlin
@SpringBootApplication
class KotlinSpringbootApplication
// 主函数，启动类
fun main(args: Array<String>) {
    runApplication<KotlinSpringbootApplication>(*args);
}
```

IndexController.kt 的代码如下：

```kotlin
@RestController
class IndexController {

    // 定义 index 接口
    @GetMapping("/index")
    fun index(): String {
        return "Hello, Kotlin for Spring Boot!!"
    }
}
```

application.yml 定义应用的配置信息：

```yaml
server:
  port: 8080 #端口号
```

通过浏览器访问"index"接口，显示"Hello，Kotlin for Spring Boot!!"。仅通过短短几行代码就开发了一个简单的 Kotlin Web 应用，非常便捷。

4.2　Kotlin 集成 Redis

Redis 是用 C 语言编写的开源的、高性能的 key-value 数据库，读取速度是 110 000 次/秒，写入速度是 81 000 次/秒，性能极高。本节介绍 Kotlin 集成 Redis 开发和 Redis 相关功能的使用知识。

4.2.1 Redis 介绍

Redis 支持数据持久化，可以将内存中的数据保存到磁盘中，重启的时候可以再次加载使用。Redis 支持丰富的数据类型，包括 string、list、set、zset、hash 等数据结构的存储。Redis 支持集群模式，master 节点可以向 slave 节点同步数据。

Redis 为单进程单线程模式，采用队列模式将并发访问变为串行访问。Redis 的所有操作都是原子性的，支持事务；Redis 还支持 publish/subscribe 通知及 key 过期等特性。

Redis 支持主从复制，当主数据库和从数据库建立主从关系后，向主数据库发送 SYNC 命令；主数据库收到 SYNC 命令后，开始在后台保存快照，并将期间收到的命令缓存；当快照完成后，主 Redis 将快照文件和所有缓存的写命令发送给从 Redis；从 Redis 接收后，会载入快照文件并且执行收到的缓存的命令；之后，每当主 Redis 收到写命令时，就将命令发送给从 Redis，从而保证数据一致。

Redis 支持高可用，通过一个或多个哨兵 Sentinel 实例组成的 Sentinel 系统可以监控任意多个主服务器，以及这些主服务器的所有从服务器。当主服务器下线时，自动将主服务器的某个从服务器升级为新的主服务器，由新的主服务器代替已下线的主服务器继续处理命令请求。Sentinel 通过心跳机制监控主从服务器工作是否正常；当主服务器出现问题时，通知相关节点；自动进行故障迁移，自动进行主从切换；支持统一的配置管理，连接者通过 Sentinel 获取主从地址。

Redis 支持的 Java 客户端有 Jedis 和 Redission 等，官方推荐使用 Redission。Jedis 是 Redis 的 Java 客户端，提供了对 Redis 命令的全面支持。Redisson 是一个高级的分布式协调 Redis 客服端，实现了分布式和可扩展的 Java 数据结构，如布隆过滤器 Bloom filter、集合 Set、ConcurrentMap 等。Redission 和 Jedis 相比，功能较为简单，不支持字符串操作，不支持排序、事务、管道、分区等 Redis 特性。

Redis 的企业级应用有如下场景。

- **分布式缓存**：缓存热点数据，可以设置过期时间，定时更新缓存数据。
- **限时业务**：Redis 的 key 可以设置过期时间，到期后 Redis 会删除这个 key。这一特性可以用于隔一段时间获取手机验证码、限时优惠活动信息等场景。
- **计数器**：Redis 的 incrby 命令可以实现原子性递增，可以用于高并发的秒杀、生成分布式序列号、接口限流等。可以限制一个手机号发送多少条短信，限制一个接口一分钟接收多少次请求，一天调用多少次等。
- **排行榜**：可使用 Redis 的 sortedSet 进行热点数据排序。例如，报表平台，可以将统计数据存入 Redis，获取排序好的数据。
- **分布式锁**：使用 Redis 的 setnx 命令，如果 key 值不存在，成功设置 key 值并缓存，同时返回 1，没有设置 key 值时返回 0。利用这个特性，在分布式环境中可以对某个对象加锁，防止多台实例同时操作。
- **点赞、好友、附近的人场景**：Redis 的 set 不允许存在重复数据，可以判断某个成员是否在集合内，可以很方便实现寻找共同好友、寻找附近的人等业务场景。
- **队列**：Redis 的 list 有 push、pop 命令，可以实现队列操作。

4.2.2 使用 Kotlin 操作 Redis

Spring Boot 集成 Redis 很简单，在 pom.xml 中添加如下依赖：

```xml
<!-- Spring Boot Redis 依赖包 -->
<dependency>
    <groupId>org.springframework.boot</groupId>
    <artifactId>spring-boot-starter-data-redis</artifactId>
</dependency>
```

在 application.yml 中添加 Redis 的配置，如 IP 地址、端口、密码等：

```yaml
spring:
  redis:
    host: 127.0.0.1    #Redis 服务器的 IP 地址
    port: 6379         #Redis 服务器的端口号
    password: 123456   #Redis 服务器密码
```

Spring Data Redis 提供操作 Redis 的类 RedisTemplate，通过注入 RedisTemplate 可以操作 Redis。RedisTemplate 提供 opsForValue、opsForList、opsForSet、opsForZset、opsForHash 分别处理单个值、list、set、zset、hash 类型。

定义 RedisData 类，注入 RedisTemplate<String, String>，key 和 value 都是 String 类型的。saveString 方法可缓存单个值；saveStringWithExpire 方法缓存单个值，且会在一定时间后过期。getString 方法可以根据 key 获取 value 值。

```kotlin
class RedisData {
    // 注入 redisTemplate
    @Autowired
```

```
4.    private lateinit var redisTemplate: RedisTemplate<String, String>
5.    // 缓存到 Redis，key 和 value 都是 String 类型的
6.    fun saveString(key: String, value: String) {
7.        redisTemplate.opsForValue().set(key, value)
8.    }
9.    // 缓存到 Redis，expireSecond 秒后过期
10.   fun saveStringWithExpire(key: String, value: String, expireSecond: Long){
11.       redisTemplate.opsForValue().set(key, value, Duration.ofSeconds(expireSecond))
12.   }
13.   // 从 Redis 获取缓存值
14.   fun getString(key: String): String? {
15.       return redisTemplate.opsForValue().get(key);
16.   }
17. }
```

saveList 方法缓存一个 List<String>，将 List 中的元素逐个推送到缓存。saveListWithExpire 会对 key 值设置缓存时间。getListValue 方法可以根据 key 值获取 List 的元素，可以指定要获取的元素的范围。

```
1.  // 将 List 缓存到 Redis
2.  fun saveList(key: String, values: List<String>) {
3.      values.forEach { v ->
4.          redisTemplate.opsForList().leftPush(key, v)
5.      }
6.  }
7.  // 将 List 缓存到 Redis，expireSecond 秒后过期
8.  fun saveListWithExpire(key: String, values: List<String>, expireSecond: Long) {
9.      values.forEach { v ->
10.         redisTemplate.opsForList().leftPush(key, v)
11.     }
12.     redisTemplate.expire(key, expireSecond, TimeUnit.SECONDS)
13. }
14. // 从 Redis 获取 List 中的元素，可以指定元素的范围
15. fun getListValue(key: String, start: Long, end: Long): List<String>? {
16.     return redisTemplate.opsForList().range(key, start, end);
17. }
```

saveSet 方法可以保存一组值，并且可以去除重复的值；saveSetWithExpire 方法可以对 key 值设置过期时间；getSetValues 方法可以根据 key 值获取所有的 value 值；getSetDiff 方

法可以根据两个集合的 key 值获取它们不同的元素的集合。

```kotlin
1.  // 将 set 缓存到 Redis,set 中没有重复元素
2.  fun saveSet(key: String, values: Array<String>) {
3.      redisTemplate.opsForSet().add(key, *values)
4.  }
5.  // 将 set 缓存到 Redis,expireSecond 秒后过期
6.  fun saveSetWithExpire(key: String, values: Array<String>, expireSecond: Long) {
7.      redisTemplate.opsForSet().add(key, *values)
8.      redisTemplate.expire(key, expireSecond, TimeUnit.SECONDS)
9.  }
10. // 从 Redis 获取缓存的 set 集合
11. fun getSetValues(key: String): Set<String>? {
12.     return redisTemplate.opsForSet().members(key)
13. }
14. // 从 Redis 获取两个 set 中不相同的元素
15. fun getSetDiff(key1: String, key2: String): Set<String>? {
16.     return redisTemplate.opsForSet().difference(key1, key2)
17. }
```

saveZset 方法可以保存一组值和值对应的得分,得分可以用于对元素进行排序。

saveZsetWithExpire 可以对 key 值设置过期时间。getZsetRangeByScore 可以获取得分在某个区间的值。

```kotlin
1.  // 将一组 pair 元素缓存到 Redis
2.  fun saveZset(key: String, values: Array<Pair<String, Double>>) {
3.      values.forEach { v -> redisTemplate.opsForZSet().add(key, v.first, v.second)}
4.  }
5.  // 将一组 pair 元素缓存到 Redis,expireSecond 秒后过期
6.  fun saveZsetWithExpire(key: String, values: Array<Pair<String, Double>>,
    expireSecond: Long) {
7.      values.forEach { v -> redisTemplate.opsForZSet().add(key, v.first, v.second)}
8.      redisTemplate.expire(key, expireSecond, TimeUnit.SECONDS)
9.  }
10. // 从 Redis 获取在某个区间的值
11. fun getZsetRangeByScore(key: String, minScore: Double, maxScore: Double):
    Set<String>? {
```

```
12.     return redisTemplate.opsForZSet().rangeByScore(key, minScore, maxScore)
13. }
```

saveHash 方法可以在 Redis 中保存一个 map 结构。saveHashWithExpire 方法可以对 key 设置过期时间。getHashValues 方法可以获取这个 map 中所有的 value 值。

```
1.  // 将一个 map 缓存到 Redis
2.  fun saveHash(key: String, values: Map<String, String>) {
3.      redisTemplate.opsForHash<String, String>().putAll(key, values);
4.  }
5.  // 将一个 map 缓存到 Redis，expireSecond 秒后过期
6.  fun saveHashWithExpire(key: String, values: Map<String, String>, expireSecond: Long) {
7.      redisTemplate.opsForHash<String, String>().putAll(key, values);
8.      redisTemplate.expire(key, expireSecond, TimeUnit.SECONDS)
9.  }
10. // 从 Redis 获取 map 中所有的 value 值
11. fun getHashValues(key: String): List<String>? {
12.     return redisTemplate.opsForHash<String, String>().values(key);
13. }
```

可以对上述方法进行测试，定义一个 RedisDataTest 类测试上述方法，注入 RedisData。通过测试可以看到，在 Redis 中缓存单个 key，如果设置过期时间，超过该时间后获取的 value 是 null。

```
1.  @SpringBootTest
2.  @RunWith(SpringRunner::class)
3.  class RedisDataTest {
4.      @Autowired
5.      lateinit var redisData: RedisData
6.      // 测试将一个字符串缓存到 Redis，2 秒后自动过期
7.      @Test
8.      fun `save redis key value with expire time`() {
9.          runBlocking {
10.             redisData.saveStringWithExpire("hello1", "helloWorld1", 2L)
11.             kotlinx.coroutines.delay(2 * 1000L)
12.         }
13.         println("get key")
```

```
14.         val value = redisData.getString("hello1")
15.         Assert.assertNull(value)
16.     }
17.     // 测试将一个值缓存到 Redis，并从缓存中取出这个值
18.     @Test
19.     fun `save and get redis value`() {
20.         val key = "hello"
21.         redisData.saveString(key, "helloWorld")
22.         val value = redisData.getString(key)
23.         Assert.assertEquals(value, "helloWorld")
24.     }
25. }
```

在 Redis 中缓存一个 List，可以获取某个范围的元素，当指定 range(0, 1)时可以获取两个元素。

```
1.  // 测试将一个 List 缓存到 Redis，2 秒后自动过期
2.  @Test
3.  fun `save redis list with expire time`() {
4.      runBlocking {
5.          val values = arrayListOf("hi1", "hi2", "hi3")
6.          redisData.saveListWithExpire("listKey1", values, 2L)
7.          kotlinx.coroutines.delay(2 * 1000L)
8.      }
9.      val values = redisData.getListValue("listKey1", 0L, 3L)
10.     Assert.assertEquals(values?.size, 0)
11. }
12. // 测试将一个 List 缓存到 Redis，并从 Redis 获取这个 List 的元素
13. @Test
14. fun `save and get list values`() {
15.     val key = "listKey"
16.     redisData.saveList(key, arrayListOf("hi1", "hi2", "hi3"))
17.     val values = redisData.getListValue("listKey", 0L, 1L)
18.     Assert.assertEquals(2, values?.size)
19. }
```

在 Redis 中保存两组元素，其中两组元素都有重复的，从下面的代码中可以看到这两

组元素中不同的元素只有 1 个。

```
1.   // 测试将一个 Set 缓存到 Redis，2s 后自动过期
2.   @Test
3.   fun `save redis set with expire time`() {
4.       runBlocking {
5.           val values = arrayOf("hello", "hello", "world")
6.           redisData.saveSetWithExpire("setKey3", values, 2L)
7.           kotlinx.coroutines.delay(2 * 1000L)
8.       }
9.       val values = redisData.getSetValues("setKey3");
10.      Assert.assertEquals(values?.size, 0)
11.  }
12.  // 测试将一个 Set 缓存到 Redis，从 Redis 取出这两个 Set 中不同的元素
13.  @Test
14.  fun `save and get redis two set diff`() {
15.      redisData.saveSet("setKey1", arrayOf("hello", "hello", "world", "wide"))
16.      redisData.saveSet("setKey2", arrayOf("hello", "hello", "world", "women"))
17.      val diffSet = redisData.getSetDiff("setKey1", "setKey2")
18.      Assert.assertEquals(diffSet?.size, 1)
19.  }
```

在 Redis 中保存一个 zset，可以指定获取某个分数范围的元素，代码如下所示，可以获取两个元素。

```
1.   // 测试将一组 pair 缓存到 Redis，2 秒后自动过期
2.   @Test
3.   fun `save redis zset with expire time`() {
4.       runBlocking {
5.           val values = arrayOf(Pair("xiaoming",98.0), Pair("xiaoli", 90.0), Pair("wangming", 100.0))
6.           redisData.saveZsetWithExpire("zsetKey2", values, 2L)
7.           kotlinx.coroutines.delay(2_000)
8.       }
9.       val values = redisData.getZsetRangeByScore("zsetKey2", 95.0, 99.0);
10.      Assert.assertEquals(values?.size, 0)
11.  }
```

```
12.  // 测试将一组 pair 缓存到 Redis，从 Redis 获取分数范围在 95.0 到 100.0 的元素
13.  @Test
14.  fun `saven and get redis zset values by score`() {
15.      redisData.saveZset("zsetKey1", arrayOf(Pair("xiaoming",98.0), Pair("xiaoli",
         90.0), Pair("wangming", 100.0)))
16.      val values = redisData.getZsetRangeByScore("zsetKey1", 95.0, 100.0)
17.      Assert.assertEquals(2, values?.size)
18.  }
```

在 Redis 中保存一个 hash，测试如下代码，可以获取 map 的 value 的数量是 2。

```
1.   // 测试将一个 map 缓存到 Redis，2 秒后自动过期
2.   @Test
3.   fun `save redis hashs with expire time`() {
4.       runBlocking {
5.           val aMap = mapOf(Pair("key1","value1"), Pair("key2", "value2"))
6.           redisData.saveHashWithExpire("hashKey2", aMap, 2L)
7.           kotlinx.coroutines.delay(2_000)
8.       }
9.       val values = redisData.getHashValues("hashKey2")
10.      Assert.assertEquals(0, values?.size)
11.  }
12.  // 测试将一个 map 缓存到 Redis，获取 map 所有的 value 值
13.  @Test
14.  fun `save and get redis hash values`() {
15.      val aMap = mapOf(Pair("key1","value1"), Pair("key2", "value2"))
16.      redisData.saveHash("hashKey1", aMap)
17.      val values = redisData.getHashValues("hashKey1")
18.      Assert.assertEquals(2, values?.size)
19.  }
```

4.3 Kotlin 集成 JPA、QueryDSL

JPA 全称是 Java Persistence API，可以通过注解描述对象和数据库表之间的关系，将实体对象持久化到数据库中。QueryDSL 弥补了 JPA 多表动态查询的不足，并且和 JPA 高度集成。本节介绍 Kotlin 集成 JPA、QueryDSL 的开发。

4.3.1 JPA、QueryDSL 介绍

JPA 提供了 ORM（对象关系映射）映射元数据，用元数据描述对象和表之间的关系，支持@Entity、@Table、@Column、@Transient 等注解。可以将开发者从烦琐的 JDBC 和 SQL 中解脱出来，使用 API 执行 CRUD 操作，框架会隐藏处理细节。JPA 是一种规范，定义了一些接口，Hibernate 是实现了 JPA 接口的 ORM 框架。Spring Data JPA 是 Spring 提供的简化 JPA 开发的框架，只需要写 DAO 层接口，就可以在不实现接口的情况下，实现对数据库的操作。其同时还支持分页、排序、复杂查询及原生 SQL 等。Spring Data JPA 是对 JPA 规范的再次封装，底层使用的还是 Hibernate 的 JPA 技术实现。MyBatis 需要在 XML 文件中定义 SQL，JPA 更加简洁。

JPA 具有如下特点。

- **标准化**：JPA 为不同数据库提供了相同的 API，这保证了基于 JPA 开发的企业应用经过少量的修改就能够在不同的 JPA 框架下运行。
- **简单易用，集成方便**：JPA 的主要目标之一是提供更加简单的编程模型，在 JPA 框架下创建实体和创建 Java 类一样简单，只需使用 javax.persistence.Entity 进行注释；JPA 的框架和接口也都非常简单。
- **可媲美 JDBC 的查询能力**：JPA 的查询语言是面向对象的，JPA 定义了独特的 JPQL，支持批量更新和修改、JOIN、GROUP BY、HAVING 等通常只有 SQL 才能够提供的高级查询特性，甚至还能支持子查询。
- **支持面向对象的高级特性**：JPA 中支持面向对象的高级特性，如类之间的继承、多态和类之间的复杂关系，最大限度地使用面向对象的模型。

对于单表查询，JPA 使用起来非常方便。但当涉及多表动态查询时，JPA 没有那么灵活，往往需要使用@Query 注解，如果在这个注解中写 SQL，那么 SQL 的可读性比较差。QueryDSL 是基于各种 ORM 上的一个通用框架，使用 QueryDSL 的 API 类库，可构建类型安全的 SQL 查询。基于 QueryDSL 可以在任何支持的 ORM 框架或者 SQL 平台上以一种通用的 API 方式来构建查询。目前，QueryDSL 支持的平台包括 JPA、JDO、SQL、Java Collections、RDF、Lucene、Hibernate Search。QueryDSL 的语法和 SQL 非常相似，代码可读性强，简洁优美。QueryDSL 使用 API 构造查询，可以安全地引用域类型和属性。

QueryDSL 支持代码自动完成，由于它基于 Java API，因此可以利用 Java IDE 的代码自动补全功能；QueryDSL 几乎可以避免所有的 SQL 语法错误；QueryDSL 采用 Domain 类型的对象和属性来构建查询，因此是类型安全的，不会因为条件类型而出现问题；QueryDSL

采用纯 Java API 构建 SQL，便于代码重构；QueryDSL 可以更轻松地定义增量查询。

QueryDSL 并不使用现有 POJO 构建查询，而是根据现有的配置生成对应的 Domain Model 构建查询。QueryDSL 插件会基于 JPA 的 POJO 实体自动生成查询实体，命名方式是 Q+对应实体名。

下面将通过具体的例子介绍如何用 Kotlin 操作 JPA 和 QueryDSL。

4.3.2 使用 Kotlin 操作 JPA、QueryDSL

Kotlin 集成 JPA、QueryDSL 需要在 pom.xml 中添加如下依赖：

```
1.  <!-- Spring Boot JPA 依赖包 -->
2.  <dependency>
3.      <groupId>org.springframework.boot</groupId>
4.      <artifactId>spring-boot-starter-data-jpa</artifactId>
5.  </dependency>
6.  <!-- QueryDSL 依赖包 -->
7.  <dependency>
8.      <groupId>com.querydsl</groupId>
9.      <artifactId>querydsl-jpa</artifactId>
10. </dependency>
11. <dependency>
12.     <groupId>com.querydsl</groupId>
13.     <artifactId>querydsl-apt</artifactId>
14. </dependency>
15. <!-- MySQL 驱动依赖包 -->
16. <dependency>
17.     <groupId>mysql</groupId>
18.     <artifactId>mysql-connector-java</artifactId>
19. </dependency>
```

此外，还需要引入 QueryDSL Maven 插件，这个插件可以根据 entity 生成 QueryDSL 的 entity。

```
1.  <!-- QueryDSL Maven 插件 -->
2.  <plugin>
3.      <groupId>com.querydsl</groupId>
```

```xml
4.         <artifactId>querydsl-maven-plugin</artifactId>
5.         <executions>
6.             <execution>
7.                 <phase>compile</phase>
8.                 <goals>
9.                     <goal>jpa-export</goal>
10.                </goals>
11.                <configuration>
12.                    <targetFolder>target/generatedsources/kotlin</targetFolder>
                       <packages>io.kang.example.entity</packages>
13.                </configuration>
14.            </execution>
15.        </executions>
16. </plugin>
```

在 application.yml 中添加如下配置，包括配置数据库的地址、用户名、密码等，jpa.hibernate.ddl-auto 为 update 可以在程序启动时根据 entity 自动在数据库中建表。

```yaml
1. spring:
2.   datasource:
3.     password: 123456    #数据库连接密码
4.     username: root      #数据库连接用户名
5.     url: jdbc:mysql://127.0.0.1:3306/video?characterEncoding=utf-8&serverTimezone=UTC   #数据库连接URL
6.   jpa:
7.     hibernate:
8.       ddl-auto: update   #如果服务启动时表格式不一致则更新表
```

定义一个 entity，名为 User。然后定义一个使用 JPA 操作 User 的接口 UserRepository，这个接口继承自 CrudRepository 接口，可根据定义的方法自动生成底层 SQL。

User.kt 的代码如下所示：

```kotlin
1. // 定义 user 实体
2. @Entity
3. data class User(
4.     // 主键，自增
5.     @Id
```

```
6.          @GeneratedValue(strategy = GenerationType.AUTO)
7.          val id: Long,
8.          val userName: String,
9.          val password: String,
10.         val email: String,
11.         val age: Int,
12.         val height: Double,
13.         val address: String,
14.         val education: EducationLevel,
15.         val income: Double
16. )
17. // 枚举类
18. enum class EducationLevel {
19.     XIAOXUE, GAOZHONG, BENKE, YANJIUSHENG, BOSHI
20. }
```

UserRepository.kt 定义了对 User 表的操作方法。CrudRepository 定义了保存单条/全部、根据主键查询、某条记录是否存在、查询全部、计数、根据主键删除、删除单条记录、批量删除、删除全部这些方法。可以自定义方法扩展操作。

```
1.  interface UserRepository: CrudRepository<User, Long> {
2.      // 根据 userName 和 password 查找用户
3.      fun findByUserNameAndPassword(userName: String, password: String): User?
        // 根据 userName 进行模糊查询
4.      fun findByUserNameLike(userName: String): List<User>?
5.      // 查找收入大于 income 的用户
6.      fun findByIncomeGreaterThan(income: Double): List<User>?
7.      // 查找用户名包含 userName 的用户
8.      fun findByUserNameContains(userName: String): List<User>?
9.      // 根据 userName 和 email 删除用户
10.     @Transactional(rollbackFor = [Exception::class])
11.     fun deleteByUserNameAndEmail(userName: String, email: String): Int?
12.     // 添加用户
13.     fun save(use: User)
14. }
```

UserRepository 提供了一些示例方法，如根据条件查找、模糊查询、查询大于某个值的记录、查询包含某个值的元素、条件删除、保存记录等。此外，可以自定义方法，在方法名上指定 SQL 操作符（插入、查询、删除等）、属性、属性的运算符等。

UserRepositoryTest.kt 对以上方法进行单元测试。testSaveUsers 方法保存三条 User 记录。

```kotlin
1.  @SpringBootTest
2.  @TestMethodOrder(MethodOrderer.OrderAnnotation::class)
3.  @ExtendWith(SpringExtension::class)
4.  class UserRepositoryTest {
5.      @Autowired
6.      lateinit var userReposiroty: UserRepository
7.      // 初始化三条测试记录
8.      @Test
9.      @Order(1)
10.     fun testSaveUsers() {
11.         userReposiroty.deleteAll()
12.         var users = arrayOf(
13.             User(0, "test01", "test01", "test01@qq.com", 45, 175.5, "Shanghai", EducationLevel.BOSHI, 50000.0),
14.             User(1, "test02", "test02", "test02@qq.com", 36, 170.5, "Shanghai", EducationLevel.YANJIUSHENG, 20000.0),
15.             User(2, "test03", "test03", "test03@qq.com", 26, 165.5, "Shanghai", EducationLevel.YANJIUSHENG, 10000.0)
16.         )
17.         userReposiroty.saveAll(users.asList().asIterable())
18.         val users1 = userReposiroty.findAll()
19.         assertEquals(3, users1.toList().size)
20.     }
```

testFindByUserNameAndPassword 方法根据用户名、密码——"test01"和"test01"——查找到一条 User 记录。

```kotlin
1.  // 测试查找用户名是"test01"、密码是"test01"的用户
2.  @Test
3.  @Order(2)
4.  fun testFindByUserNameAndPassword() {
5.      val user = userReposiroty.findByUserNameAndPassword("test01", "test01")
6.      assertEquals("test01@qq.com", user?.email)
```

```
7.     assertEquals(45, user?.age)
8.     assertEquals(175.5, user?.height)
9.     assertEquals(EducationLevel.BOSHI, user?.education)
10. }
```

testFindByUserNameLike 方法根据用户名进行模糊查询，查询用户名中包含"test"字符串的记录。

```
1. // 测试查找用户名包含"test"字符串的用户
2. @Test
3. @Order(3)
4. fun testFindByUserNameLike() {
5.     val users = userReposiroty.findByUserNameLike("%test%")
6.     assertEquals(3, users?.size)
7. }
```

testFindByIncomeGreaterThan 方法查询收入大于 10 000.0 的记录，此外，还支持大于或等于、等于、小于、小于或等于这些操作符。

```
1. // 测试查找收入大于 10 000.0 的用户
2. @Test
3. @Order(4)
4. fun testFindByIncomeGreaterThan() {
5.     val users = userReposiroty.findByIncomeGreaterThan(10000.0)
6.     assertEquals(2, users?.size)
7. }
```

testFindByUserNameContains 方法查找用户名包含"test"字符串的记录。

```
1. // 测试查找用户名包含"test"字符串的用户
2. @Test
3. @Order(5)
4. fun testFindByUserNameContains() {
5.     val users = userReposiroty.findByUserNameContains("test")
6.     assertEquals(3, users?.size)
7. }
```

testDeleteByUserNameAndEmail 方法根据用户名和邮箱删除记录，删除用户名是

"test01"、邮箱是"test01@qq.com"的记录。

```
1.  // 测试根据用户名和邮箱删除用户
2.  @Test
3.  @Order(6)
4.  fun testDeleteByUserNameAndEmail() {
5.      userReposiroty.deleteByUserNameAndEmail("test01", "test01@qq.com")
6.      val users = userReposiroty.findAll().toList()
7.      assertEquals(2, users.size)
8.  }
```

testSave 方法保存单条记录。

```
1.  // 测试保存一条用户记录
2.  @Test
3.  @Order(7)
4.  fun testSave() {
5.      val user = User(3, "test04", "test04", "test04@qq.com", 26, 165.5, "Shanghai", EducationLevel.YANJIUSHENG, 10000.0)
6.      userReposiroty.save(user)
7.      val users = userReposiroty.findAll().toList()
8.      assertEquals(3, users.size)
9.  }
```

注入 JPAQueryFactory,可以使用 QueryDSL 操作数据库。用变量 predicate 保存 where 条件,queryFactory 提供查询、删除和更新的方法。JPAQueryFactory 操作的是 QUser 对象,是使用插件在 User 基础上生成的。

testFindByUserNameAndPassword 方法展示了使用 queryDSL 查询用户名和密码分别是"test01"和"test01"的记录。

```
1.  @SpringBootTest
2.  @TestMethodOrder(MethodOrderer.OrderAnnotation::class)
3.  @ExtendWith(SpringExtension::class)
4.  class UserDomainRepositoryTest {
5.      // 注入 queryFactory
6.      @Autowired
7.      lateinit var queryFactory: JPAQueryFactory
```

```
8.      // 测试根据用户名、密码查找用户
9.      @Test
10.     @Order(1)
11.     fun testFindByUserNameAndPassword() {
12.         val qUser = QUser.user
13.         val predicate = qUser.userName.eq("test01").and(qUser.password.eq("test01"))
14.         val user = queryFactory.selectFrom(qUser).where(predicate).fetchOne()
15.         Assert.assertEquals("test01@qq.com", user?.email)
16.         Assert.assertEquals(45, user?.age)
17.         Assert.assertEquals(175.5, user?.height)
18.         Assert.assertEquals(EducationLevel.BOSHI, user?.education)
19.     }
20. }
```

testFindByUserNameLike 方法展示了使用 queryDSL 对用户名进行模糊查询。

```
1.  // 测试用户名包含 "test" 字符串的用户
2.  @Test
3.  @Order(2)
4.  fun testFindByUserNameLike() {
5.      val qUser = QUser.user
6.      val predicate = qUser.userName.like("%test%")
7.      val users = queryFactory.selectFrom(qUser).where(predicate).fetch()
8.      Assert.assertEquals(3, users?.size)
9.  }
```

testFindByIncomeGreaterThan 方法展示了使用 queryDSL 查询收入大于 10 000.0 的记录。

```
1.  // 测试收入大于 10 000.0 的用户
2.  @Test
3.  @Order(3)
4.  fun testFindByIncomeGreaterThan() {
5.      val qUser = QUser.user
6.      val predicate = qUser.income.gt(10000.0)
7.      val users = queryFactory.selectFrom(qUser).where(predicate).fetch()
8.      Assert.assertEquals(2, users.size)
9.  }
```

testFindByUserNameContains 方法展示了使用 queryDSL 查询用户名包含 "test" 的记录。

```kotlin
1.  // 测试用户名包含"test"的用户
2.  @Test
3.  @Order(4)
4.  fun testFindByUserNameContains() {
5.      val qUser = QUser.user
6.      val predicate = qUser.userName.contains("test")
7.      val users = queryFactory.selectFrom(qUser).where(predicate).fetch()
8.      Assert.assertEquals(3, users.size)
9.  }
```

testDeleteByUserNameAndEmail 方法展示了使用 queryDSL 删除用户名和邮箱分别是"test01"和"test01@qq.com"的记录。由于涉及删除操作，因此需要事务注解。

```kotlin
1.  // 测试根据用户名和邮箱删除用户
2.  @Test
3.  @Order(5)
4.  @Transactional
5.  @Rollback(false)
6.  fun testDeleteByUserNameAndEmail() {
7.      val qUser = QUser.user
8.      val predicate = qUser.userName.eq("test01").and(qUser.email.eq("test01@qq.com"))
9.      queryFactory.delete(qUser).where(predicate).execute()
10.     val users = queryFactory.selectFrom(qUser).fetch()
11.     Assert.assertEquals(2, users.size)
12. }
```

testUpdateEmailByUserName 方法展示了使用 queryDSL 把用户名是"test02"的记录的邮箱更新为"test02@yy.com"。

```kotlin
1.  // 测试更新用户名是 test02 的用户的邮箱
2.  @Test
3.  @Order(6)
4.  @Transactional
5.  @Rollback(false)
6.  fun testUpdateEmailByUserName() {
7.      val qUser = QUser.user
8.      val predicate = qUser.userName.eq("test02")
```

```
9.      queryFactory.update(qUser).set(qUser.email,
"test02@yy.com").where(predicate).execute()
10.     val user = queryFactory.selectFrom(qUser).where(predicate).fetchOne()
11.     Assert.assertEquals("test02@yy.com", user?.email)
12. }
```

4.4 Kotlin 集成 MongoDB

MongoDB 是用 C++编写的高性能非关系数据库,是基于分布式文件存储的开源数据库系统。MongoDB 为企业级应用提供了可扩展的数据存储解决方案。本节介绍 MongoDB 的背景知识以及如何使用 Kotlin 集成 MongoDB 开发。

4.4.1 MongoDB 介绍

MongoDB 将数据存储为一个文档,数据结构由键值对组成,类似 JSON 对象。字段值可以包含 null、布尔值、整数、浮点数、字符串、唯一 id、日期、正则表达式、代码、undefined、数组及其他文档。

MongoDB 是一个面向文档的数据库,操作起来比较简单。它可以对任何属性建立索引,以提高检索速度。它还支持丰富的查询表达式,查询指令使用 JSON 形式的标记,可轻易查询文档中内嵌的对象以及数组。它使用 update 命令更新文档的数据。它支持 map/reduce 操作,可对数据进行批量处理和聚合操作。它使用 map 函数调用 emit(key, value)遍历集合中所有的记录,将 key 和 value 传给 reduce 函数进行处理。mongoDB 允许在服务端执行脚本,用 JavaScript 编写函数,直接在服务端运行,也可以把函数的定义存储在服务端,客户端直接调用该函数。MongoDB 支持 Ruby、Python、Java、Kotlin、C++、PHP、C#等编程语言。

一台 MongoDB 的存储容量有限,当需要扩展存储容量时,可以进行水平扩展。MongoDB 通过 Sharded cluster 保证可扩展性,Sharded 是指复制集或者单个 Mongos 节点。Sharded cluster 由 Shard、Mongos 和 Config server 等 3 个组件构成。Mongos 是 Sharded cluster 的访问入口。Mongos 本身并不持久化数据,Sharded cluster 所有的元数据都会存储到 Config server,而用户的数据则会分散存储到各个 Shard。Mongos 启动后,会从 Config server 加载元数据,开始提供服务,将用户的请求正确路由到对应的 Shard。Sharded cluster 支持将单个集合的数据分散存储在多个 Shard 上,用户可以指定根据集合内文档的某个字段即 shard key 来分布数据,目前主要支持两种数据分布策略——范围分片(Range based sharding)或

hash 分片（Hash based sharding）。Mongos 会根据请求类型及 shard key 将请求路由到对应的 Shard。Config server 存储 Sharded cluster 的所有元数据，所有的元数据都存储在 Config 数据库中。config.shards 集合存储各个 Shard 的信息，可通过 addShard、removeShard 命令动态地从 Sharded cluster 里增加或移除 Shard。config.databases 集合存储所有数据库的信息，包括数据库是否开启分片和 primary shard 的信息等，对于数据库内没有开启分片的集合，所有的数据都会存储在数据库的 primary shard 上。

  MongoDB 4.0 以上版本支持事务，MongoDB 4.0 支持副本集事务，存在单个文档最大 16MB、事务执行时间不能过长的限制。MongoDB 4.2 版本支持分布式事务，支持修改分片 key 的内容。事务的执行过程是：获取会话，开启事务，获得集合，多个集合操作，回滚事务，提交事务。

  MongoDB 适用于存储日志，不需要像关系数据库那样设计统一的表结构，存储更加灵活方便。存储监控数据、增加字段都不需要修改表结构，使用成本低。一些 O2O 快递应用，将骑手、商家、位置信息保存在 MongoDB 中，通过地理位置查询附近的商家、骑手等。在证券交易类应用中，将用户的交易数据存储到 MongoDB 中，根据这些交易数据可定时生成市场行情。

4.4.2　使用 Kotlin 操作 MongoDB

  Kotlin 集成 MongoDB 需要在 pom.xml 中添加如下依赖：

```xml
1. <!-- Spring Boot MongoDB 依赖包 -->
2. <dependency>
3.     <groupId>org.springframework.boot</groupId>
4.     <artifactId>spring-boot-starter-data-mongodb</artifactId>
5. </dependency>
```

在 application.yml 中添加如下配置，配置 MongoDB 的 URL，我们使用的是单机 MongoDB：

```yaml
1. spring:
2.   data:
3.     mongodb:
4.       uri: mongodb://exchange:123456@127.0.0.1:27017/exchange#mongodb 连接 URL
```

  Person.kt 定义了集合和实体的映射关系，Student 继承了 Person，Student 集合具有 personId、name、address、age、date、likeSport、likeBook、school 这些属性。

```kotlin
1.  // 定义 Person 实体类
2.  open class Person(
3.          @Id
4.          var personId: Long = 0L,
5.          var name: String = "",
6.          var address: String = "",
7.          var age: Int = 0,
8.          var date: Date = Date()
9.  )
10. // 定义一个 Student 实体类，映射为 MongoDB 中的 Student 集合
11. @Document(collection="student")
12. data class Student(
13.         val likeSport: String,
14.         val likeBook: String,
15.         val school: String
16. ): Person()
```

StudentRepository.kt 定义了操作 Student 集合的接口，它继承了 MongoRepository。MongoRepository 提供了批量/单笔保存、全部查询、条件查询等基本方法。它还可以扩展集合的操作方法，方法名由操作方法、属性和操作符组成；也可以使用@DeleteQuery、@Query 注解自定义原生的删除、查询语句。

```kotlin
1.  interface StudentRepository: MongoRepository<Student, Long> {
2.      // 根据 personId 查找 student
3.      fun findByPersonId(personId: Long): Student?
4.      // 分页查找年龄大于或等于 age，模糊匹配 name 的 student
5.      fun findByNameRegexAndAgeGreaterThanEqual(name: String, age: Int, pageable: Pageable): Page<Student>
6.      // 使用原生 SQL 删除年龄在 age1 到 age2 范围的数据
7.      @DeleteQuery(value = "{\"age\":{\"$gte\":?0,\"$lte\":?1}}")
8.      fun deleteByAgeIn(age1: Int, age2: Int)
9.      // 使用原生 SQL 分页查找年龄大于或等于 age，模糊匹配 name 的 student
10.     @Query(value = "{\"name\":{\"$regex\":?0},\"age\":{\"$gte\":?1}}")
11.     fun findByAgeIndividual(name: String ,age: Int, pageable: Pageable): Page<Student>
12. }
```

StudentRepositoryTest.kt 对 StudentRepository 定义的方法进行了测试。此外，也可以使

用 MongoTemplate 来操作集合。saveStudents 方法向 Student 集合中保存了 16 条记录。

```kotlin
1.  @SpringBootTest
2.  @TestMethodOrder(MethodOrderer.OrderAnnotation::class)
3.  @ExtendWith(SpringExtension::class)
4.  class StudentRepositoryTest {
5.      @Autowired
6.      lateinit var studentRepository: StudentRepository
7.      @Autowired
8.      lateinit var mongoTemplate: MongoTemplate
9.      // 测试保存 student 数据,初始化 16 条测试记录
10.     @Test
11.     @Order(1)
12.     fun saveStudents() {
13.         val student = Student("篮球","Kotlin 编程","上海中学")
14.         student.personId = 20180101L
15.         student.age = 22
16.         student.name = "张三"
17.
18.         studentRepository.deleteAll()
19.
20.         for(i in 0..15) {
21.             student.age = 22 - i
22.             student.personId = 20180101L + i
23.             studentRepository.save(student)
24.         }
25.     }
26. }
```

testFindByNameAndAge 方法对名字是"张三"、年龄大于或等于 20 的记录进行分页查询,每页 5 条记录,查询第 0 页。

```kotlin
1.  // 测试查找名字是"张三"、年龄大于或等于 20 的 student,查到 3 条
2.  @Test
3.  @Order(2)
4.  fun testFindByNameAndAge() {
5.      val pageable = PageRequest.of(0, 5)
```

```
6.
7.     val studentPage = studentRepository.findByNameRegexAndAgeGreaterThanEqual
("张三", 20, pageable)
8.
9.     Assert.assertEquals(3, studentPage.content.size)
10. }
```

testFindByNameAndAge1 方法使用原生的查询语句对姓名是"张三"、年龄大于或等于 10 的记录进行分页查询。用 JSON 格式定义查询条件：{"name":{"$regex":"张三"},"age":{"$gte":10}}，而 testFindByNameAndAge 是 Mongo JPA 根据定义的方法自动生成查询语句。

```
1. // 测试使用原生 SQL 查找名字是"张三"、年龄大于或等于 10 的 student，查到 5 条
2. @Test
3. @Order(3)
4. fun testFindByNameAndAge1() {
5.     val pageable = PageRequest.of(0, 5)
6.     val studentPage = studentRepository.findByAgeIndividual("张三", 10, pageable)
7.     Assert.assertEquals(5, studentPage.content.size)
8. }
```

testDeleteByAgeIn 方法删除年龄大于或等于 20 且小于或等于 22 的记录。

```
1. // 测试删除年龄大于或等于 20 且小于或等于 22 的 student
2. @Test
3. @Order(4)
4. fun testDeleteByAgeIn() {
5.     studentRepository.deleteByAgeIn(20, 22)
6.     val students = studentRepository.findAll()
7.     Assert.assertEquals(13, students.size)
8. }
```

testSortByAge 方法使用 mongoTemplate 操作集合，通过 aggregation 定义对集合的具体操作：指定操作的集合，根据年龄降序查找，只取 1 条记录。

```
1. // 测试使用 mongoTemplate 根据 age 降序查找，只取 1 条记录
2. @Test
3. @Order(5)
```

```
4.  fun testSortByAge() {
5.      val aggregation = Aggregation.newAggregation(Student::class.java,
    sort(Sort.Direction.DESC, "age"), limit(1))
6.      val student = mongoTemplate.aggregate(aggregation,
    Student::class.java).mappedResults
7.      Assert.assertEquals(19, student[0].age)
8.  }
```

testFindCount 方法查询集合 Student 的总数。

```
1.  // 测试使用 mongoTemplate 查找 Student 集合的总数
2.  @Test
3.  @Order(6)
4.  fun testFindCount() {
5.      val n = mongoTemplate.db.getCollection("student").countDocuments()
6.      Assert.assertEquals(13, n)
7.  }
```

testFindByName 方法先获取集合 Student，返回满足姓名等于"张三"、年龄大于或等于 19 的记录的总数。

```
1.  // 测试使用 mongoTemplate 查找名字是"张三"、年龄大于或等于 19 的 student 的总数
2.  @Test
3.  @Order(7)
4.  fun testFindByName() {
5.      val n = mongoTemplate.db.getCollection("student").find()
6.          .filter(eq("name", "张三"))
7.          .filter(gte("age", 19)).count()
8.      Assert.assertEquals(1, n)
9.  }
```

testFindByNameLimit 方法获取集合 Student，过滤姓名等于"张三"的记录，取前 5 条，并返回总数。

```
1.  // 测试使用 mongoTemplate 查找名字是"张三"的记录，取 5 条记录
2.  @Test
3.  @Order(8)
4.  fun testFindByNameLimit() {
```

```
5.      val n = mongoTemplate.db.getCollection("student").find()
6.              .filter(eq("name", "张三"))
7.              .limit(5)
8.              .count()
9.      Assert.assertEquals(5, n)
10. }
```

testFindByNameCurse 方法对集合 Student 进行操作，过滤姓名等于"张三"的记录，打开一个游标，通过游标可以遍历这些记录。

```
1.  // 测试使用 mongoTemplate 查找 name 是"张三"的记录，并用游标遍历元素
2.  @Test
3.  @Order(9)
4.  fun testFindByNameCurse() {
5.      val studentCursor = mongoTemplate.db.getCollection("student").find()
6.              .filter(eq("name", "张三"))
7.              .cursor()
8.      while(studentCursor.hasNext()){
9.          Assert.assertEquals("张三", studentCursor.next()["name"])
10.     }
11.     studentCursor.close()
12. }
```

testFindObjectId 方法重点测试 MongoDB 的主键_id，_id 的类型是 ObjectId。ObjectId 是一个分布式 id，使用 12 字节的存储空间，是一个 24 位的字符串。前 4 字节是一个时间戳，单位是秒，表示文档的创建时间；接下来的 3 字节是所在主机的唯一标识符；接下来的 2 字节来自产生 ObjectId 的进程标识符；后 3 字节是一个自动增加的计数器，确保相同进程同一秒产生的 ObjectId 是不一样的。可以利用 ObjectId 的时间戳对整个集合进行分段、快速遍历。当集合文档有上亿条时，采用分页查询，越往后越慢。此时如果巧妙利用 ObjectId 的时间戳，可以将集合根据时间分段，多线程地去遍历集合。

```
1.  // 测试使用 mongoTemplate 查找 name 是"张三"的记录，打印主键_id
2.  @Test
3.  @Order(10)
4.  fun testFindObjectId() {
5.      val studentCursor = mongoTemplate.db.getCollection("student").find()
6.              .filter(eq("name", "张三"))
```

```
7.            .cursor()
8.        while(studentCursor.hasNext()){
9.            val document = studentCursor.next()
10.           println("_id: ${document["_id"]}, date: ${document["date"]}")
11.       }
12.       studentCursor.close()
13.   }
```

testDelete 方法使用 mongoTemplate 删除集合的文档，使用 document 定义删除的条件，将姓名是"张三"的记录全部删除。

```
1.    // 测试使用 mongoTemplate 删除 name 是"张三"的所有记录
2.    @Test
3.    @Order(11)
4.    fun testDelete() {
5.        val document = Document()
6.        document["name"] = "张三"
7.        mongoTemplate.db.getCollection("student").deleteMany(document)
8.        val n= mongoTemplate.db.getCollection("student").countDocuments()
9.        Assert.assertEquals(0, n)
10.   }
```

4.5 Kotlin 集成 Spring Security

Spring Security 是为基于 Spring 的企业应用系统提供声明式的安全访问控制的解决方案。它利用 Spring 的依赖注入和面向切面技术，提供了声明式的安全访问控制功能，可以在 Web 请求级别和方法调用级别处理身份认证和授权。本节主要介绍 Kotlin 集成 Spring Security 开发。

4.5.1 Spring Security 介绍

身份认证是指验证某个用户是否是系统中的合法用户，一般要求用户通过登录界面填写用户名、密码及验证码等信息。系统通过校验用户名和密码完成认证。用户授权是指某个用户是否有权限执行某个操作。系统会为不同的用户分配不同的角色，每个角色对应一系列的权限。Spring Security 在进行用户认证以及授予权限的时候，通过各种各样的拦截器来控制权限的访问,从而实现系统安全访问接口。

Spring Security 的主要模块有如下几个。

- ACL:支持通过访问控制列表为域对象提供安全性。
- Aspects:当使用 Spring Security 注解时,会使用基于 AspectJ 的切面。
- CAS Client:提供与 CAS 集成的功能。
- Configuration:包含对 XML、Java 配置功能的支持。
- Core:基本的功能库。
- Cryptography:提供加密和解密编码相关功能。
- LDAP:提供基于 LDAP 的认证。
- OpenID:支持使用 OpenID 进行集中式认证。
- Web:提供基于 Filter 的 Web 安全支持。

在用户认证方面,Spring Security 支持主流的认证方式——HTTP 基本认证、HTTP 表单验证、HTTP 摘要认证、OPENID 和 LDAP 等。Spring Security 支持基于内存的验证、基于数据库的验证及用户自定义服务的验证。用户一次完整的登录验证和授权,是一个请求经过层层拦截器从而实现权限控制的过程。整个 Web 端配置为 DelegatingFilterProxy,它是所有过滤器链的代理类,真正执行拦截处理的是由 Spring 容器管理的 Filter Bean 组成的 FilterChain。用户认证涉及的主要类有 UserDetails,其作为整个登录过程的 POJO 接口;UserDetailsService 用于生成 UserDetails;UsernamePasswordAuthenticationFilter,使用初始化中的 AuthenticationManager 的 AbstractUserDetailsAuthenticationProvider 的 authenticate 方法进行校验;AbstractUserDetailsAuthenticationProvider,首先会通过 retrieveUser 方法使用我们定义的 UserDetailsService 生成对应的 UserDetails,获得 UserDetails 后,再使用自身的 additionalAuthenticationChecks 方法去验证数据库的用户信息和登录信息是否一致;如果 additionalAuthenticationChecks 没有报错,那么请求就会带着 UserDetails 的权限成功登录。

在用户授权方面,Spring Security 提供基于角色的访问控制和访问控制列表。Spring Security 有三种不同的安全注解:自带的@Secured 注解,@EnableGlobalMethodSecurity 中设置 securedEnabled = true;JSR-250 的@RolesAllowed 注解,@EnableGlobalMethodSecurity 中设置 jsr250Enabled = true;表达式驱动的注解。@PreAuthorize 方法调用之前,基于表达式的计算结果限制对方法的访问;@PostAuthorize 方法调用之后,如果表达式的结果为 false,则抛出异常。@PreFilter 方法调用之前,过滤进入方法的输入值;@PostFilter 方法调用之后,过滤方法的结果值。

Spring Security 在很多企业级应用中用于后台角色权限控制、授权认证、安全防护(防止跨站点请求）、防止 Session 攻击，非常容易在 Spring MVC 中使用。

4.5.2 使用 Kotlin 操作 Spring Security

Kotlin 集成 Spring Security 需要在 pom.xml 中添加如下依赖：

```xml
<!-- Spring Boot Security 依赖包 -->
<dependency>
    <groupId>org.springframework.boot</groupId>
    <artifactId>spring-boot-starter-security</artifactId>
</dependency>
```

WebSecurityConfig.kt 继承自 WebSecurityConfigurerAdapter，重载 configure(HttpSecurity) 方法，配置如何通过拦截器保护请求。

```kotlin
@Configuration
@EnableWebSecurity//启用安全
@EnableGlobalMethodSecurity(prePostEnabled = true)//开启注解
class WebSecurityConfig: WebSecurityConfigurerAdapter() {
    //所有的接口都要进行授权才能访问，关闭跨域保护
    override fun configure(http: HttpSecurity) {
        http.formLogin()
                .permitAll()
                .and()
                .authorizeRequests()
                .antMatchers("/**")
                .authenticated()
                .and()
                .csrf().disable()
    }
}
```

CustomAuthProvider.kt 继承自 AuthenticationProvider，重载 authenticate(Authentication) 方法，对用户权限进行校验。根据登录用户名、密码去数据库查询，如果能查询到，则用户登录成功，再根据用户名查找相应的权限。

```kotlin
@Component
class CustomAuthProvider: AuthenticationProvider {
```

```
3.      // 注入 UserRepository
4.      @Autowired
5.      lateinit var userRepository: UserRepository
6.      // 用户、角色列表
7.      private val auth = mapOf(Pair("test03", "ROLE_USER"), Pair("test02",
   "ROLE_ADMIN"))
8.      // 授权方法
9.      @Throws(AuthenticationException::class)
10.     override fun authenticate(authentication: Authentication): Authentication? {
11.         // 用户名、密码
12.         val username = authentication.name
13.         val password = authentication.credentials.toString()
14.         // 使用用户名、密码查找是否有这个用户
15.         val user = userRepository.findByUserName(username)
16.         // 获取该用户的角色
17.         if(user?.password.equals(password)) {
18.             val authorities = AuthorityUtils.commaSeparatedStringToAuthorityList
   (auth[username])
19.             return UsernamePasswordAuthenticationToken(user, password,
   authorities)
20.         }
21.         return null
22.     }
23.     override fun supports(p0: Class<*>?): Boolean {
24.         return true;
25.     }
26. }
```

IndexController.kt 定义了几个接口和相应的权限校验。

```
1. @RestController
2. class IndexController {
3.     // 测试接口，任何角色的用户都可以访问
4.     @GetMapping("/")
5.     fun index(): String {
6.         return "Hello, Kotlin for Spring Boot!!"
7.     }
```

```kotlin
8.      // 测试接口，USER 角色的用户可以访问，在访问方法前进行角色校验
9.      @PreAuthorize("hasRole('USER')")
10.     @GetMapping("/hello/pre/user")
11.     fun rolePreUserHello(): String {
12.         println("pre filter admin user")
13.         return "Hello, I have admin role"
14.     }
15.     // 测试接口，USER 角色的用户可以访问，在访问方法后进行角色校验
16.     @PostAuthorize("hasRole('USER')")
17.     @GetMapping("/hello/post/user")
18.     fun rolePostUserHello(): String {
19.         println("post filter user role")
20.         return "Hello, I have user role"
21.     }
22.     // 测试接口，ADMIN 角色的用户可以访问，在访问方法前进行角色校验
23.     @PreAuthorize("hasRole('ADMIN')")
24.     @GetMapping("/hello/admin")
25.     fun roleAdminHello(): String {
26.         println("pre filter admin user")
27.         return "Hello, I have admin role"
28.     }
29.     // 测试接口，ADMIN、USER 角色的用户可以访问，在访问方法前进行角色校验
30.     @PreAuthorize("hasAnyRole('USER', 'ADMIN')")
31.     @GetMapping("/hello/any")
32.     fun anyRoleUserHello(): String {
33.         return "Hello, I have one of [user, admin] role"
34.     }
35.     // 测试接口，USER 角色的用户可以访问，在访问方法前进行角色校验
36.     // 只传递年龄大于 50 的 user 的数据
37.     @PreFilter(value="hasRole('USER') and filterObject.age > 50")
38.     @PostMapping("/user/prefilter")
39.     fun preFilterUser(@RequestBody user: List<User>): List<User> {
40.         println("pre filter user")
41.         return user
42.     }
43.     // 测试接口，USER 角色的用户可以访问，在访问方法后进行角色校验
44.     // 只传递年龄大于 50 的 user 的数据
```

```
45.    @PostFilter(value="hasRole('USER') and filterObject.age > 50")
46.    @PostMapping("/user/postfilter")
47.    fun postFilterUser(@RequestBody user: List<User>): List<User> {
48.        println("post filter user")
49.        return user
50.    }
51.    // 测试接口，ADMIN 角色的用户可以访问，在访问方法前进行角色校验
52.    // 只传递年龄大于 50 的 user 的数据
53.    @PreFilter(value="hasRole('ADMIN') and filterObject.age > 50")
54.    @PostMapping("/user/admin/prefilter")
55.    fun preFilterAdmin(@RequestBody user: List<User>): List<User> {
56.        println("pre filter user")
57.        return user
58.    }
59.    // 测试接口，USER 角色的用户可以访问，在访问方法前进行角色校验
60.    // 只传递 userName 等于 test02 的 user 的数据
61.    @PreFilter(value="hasRole('ADMIN') and filterObject.userName.equals('test02')")
62.    @PostMapping("/user/prefilter1")
63.    fun preFilterAdmin1(@RequestBody user: List<User>): List<User> {
64.        println("pre filter user")
65.        return user
66.    }
67. }
```

输入用户名 test02、密码 test02 登录后显示 Hello, Kotlin for Spring Boot!!，test02 用户具有 admin 角色。

rolePreUserHello 方法用 @preAuthorize 校验用户是否具有 "USER" 角色。rolePostUserHello 方法用 @postAuthorize 校验用户是否具有 "USER" 角色。@preAuthorize 在方法调用之前，基于表达式的计算结果来限制对方法的访问；@PostAuthorize 在方法调用之后，如果表达式的结果为 false，则抛出异常。调用 rolePostUserHello 方法，虽然 test02 用户没有 "USER" 角色，仍然打印 "post filter user role"。

roleAdminHello 方法使用 @preAuthorize 校验用户是否具有 "ADMIN" 角色。anyRoleUserHello 方法校验用户是否具有 "ADMIN" 或者 "USER" 角色。test02 用户具有 "USER" 角色，会正常访问 anyRoleUserHello 方法，没有权限访问 roleAdminHello 方法。

图4.2 /user/prefilter接口返回参数

preFilterUser 方法使用@PreFilter 校验用户是否具有"USER"角色，并对输入的参数进行过滤，把年龄大于 50 的 user 对象传入。postFilterUser 方法使用@PostFilter 校验用户是否具有"USER"角色，对返回结果进行过滤。@PreFilter 在方法调用之前，过滤进入方法的输入值；@PostFilter 在方法调用之后，过滤方法的结果值。两个方法都返回空，如图 4.2 所示。

preFilterAdmin1 方法使用@PreFilter 校验用户是否具有"ADMIN"角色，并对输入的参数进行过滤，把 userName 等于"test02"的 user 对象传入，如图 4.3 所示。

图4.3 /user/prefilter1接口返回参数

preFilterAdmin 方法使用 @ PreFilter 校验用户是否具有"ADMIN"角色，并对输入的参数进行过滤，把年龄大于 50 的 user 对象传入，如图 4.4 所示。

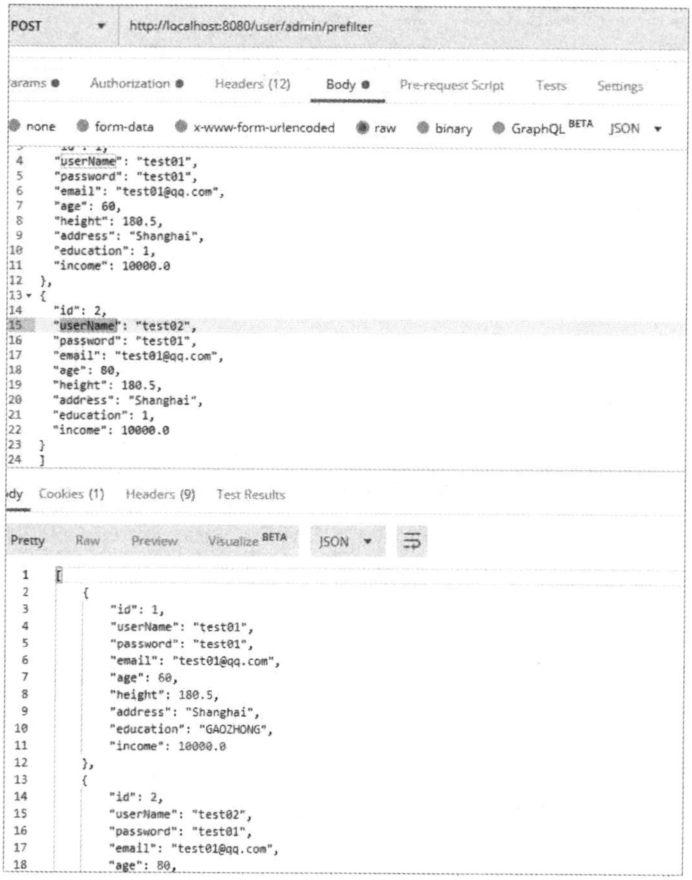

图4.4 /admin/prefilter接口返回参数

4.6 Kotlin 集成 RocketMQ

RocketMQ 是一款用 Java 语言编写的分布式的队列模型的开源消息中间件，支持事务消息、顺序消息、批量消息、定时消息及消息回溯等。本节介绍 Kotlin 集成 RocketMQ 开发。

4.6.1 RocketMQ 介绍

RocketMQ 由阿里巴巴研发，现在是 Apache 顶级项目。其采用 Netty NIOl 框架实现数据通信，用轻量级的 NameServer 进行网络路由，提供了服务性能，并支持消息失败重试机制。它支持集群模式、消费者负载均衡、水平扩展能力并支持广播模式。其采用零拷贝原理，顺序写盘、支持亿级消息堆积能力。

RocketMQ 架构分为四个模块：生产者，消费者，NameServer，BrokerServer。组件均设计为无状态，任何组件的节点都支持集群部署扩展。

生产者充当消息发布者的角色，支持分布式集群方式部署。生产者通过 MQ 的负载均衡模块选择相应的 Broker 集群队列进行消息投递，投递的过程支持快速失败并且低延迟。

消费者充当消息消费者的角色，支持分布式集群方式部署。支持以 push 和 pull 两种模式对消息进行消费。同时也支持集群方式和广播形式的消费，它提供实时消息订阅机制，可以满足大多数用户的需求。

NameServer 是基于服务注册发现功能的无状态组件，支持独立部署。NameServer 接收 Broker 集群的注册信息并且将其保存下来作为路由信息的基本数据，然后提供心跳检测机制，检查 Broker 是否还存活。每个 NameServer 保存关于 Broker 集群的整个路由信息和用于客户端查询的队列信息，然后生产者和消费者通过 NameServer 就可以知道整个 Broker 集群的路由信息，从而进行消息的投递和消费。

Broker 是基于高性能和低延迟的文件存储的无状态组件，支持独立部署。主要负责消息的存储、投递和查询以及保证服务的高可用性。Broker 包含如下模块。

- Remoting Module：整个 Broker 的实体，负责处理来自客户端的请求。
- Client Manager：负责管理客户端（生产者/消费者）和维护消费者的 topic 订阅信息。
- Store Service：提供简单方便的 API 接口，提供将消息存储到物理硬盘和查询的功能。
- HA Service：高可用服务，提供 master Broker 和 slave Broker 之间的数据同步功能。
- Index Service：根据特定的 Message key 对投递到 Broker 的消息进行索引服务，以提供消息的快速查询。

RocketMQ 支持集群部署，可提高系统的吞吐量和可用性。namesrv 可以独立部署，且 namesrv 与 namesrv 之间并无直接或间接的关联，双方不存在心跳检测，所以 namesrv 之间不存在主从切换过程，如果其中一台 namesrv 宕机，生产者和消费者会直接从另一台 namesrv 中请求数据。Broker 支持主从架构模式，目前它支持 master 写操作，只有当 master 读压力高于某个点（master 消息拉取出现堆积）时，才会将读压力转给 slaver。无法做到主从切换，master 宕机，slaver 只能提供消息消费，slaver 不会被选举为 master 来继续工作。如果 master

宕机，整个消息队列环境近乎瘫痪。如果用户体量稍微大一些，单 master/单 slaver 架构扛不住，可以采用多 master/多 slaver 部署架构。

RocketMQ 提供了两种刷盘方式，即同步刷盘和异步刷盘。同步刷盘指消息投放到 Broker 之后，会在写入文件之后才返回成功，同步刷盘可保证数据不丢失。异步刷盘则指消息投放到 Broker 成功后即可返回，同时启动另外的线程来存储消息。

RocketMQ 可以应用于异步解耦、消息的顺序收发、削峰填谷、分布式事务数据的一致性及大规模机器的缓存同步等业务场景。

4.6.2 使用 Kotlin 操作 RocketMQ

Kotlin 集成 RocketMQ 需要在 pom.xml 中添加如下依赖：

```
1. <!-- Spring Boot RocketMQ 依赖包 -->
2. <dependency>
3.     <groupId>org.apache.rocketmq</groupId>
4.     <artifactId>rocketmq-spring-boot-starter</artifactId>
5.     <version>2.0.4</version>
6. </dependency>
```

在 application.yml 中添加如下 RocketMQ 的配置，我们使用的是单机 RocketMQ：

```
1. rocketmq:
2.   name-server: 127.0.0.1:9876   #RocketMQ 的主机、端口
3.   producer:
4.     group: kotlin-group   #生产者 group
5.     send-message-timeout: 300000   #发送消息的超时时间
6.     compress-message-body-threshold: 4096   #压缩消息的阈值 4KB
7.     max-message-size: 4194304   #消息最大 4MB
8.     retry-times-when-send-failed: 2   #发送失败后重试 2 次
9.     retry-next-server: true   #开启内部消息重试
10.    retry-times-when-send-async-failed: 0   #异步发送重试 0 次
```

OrderPaidEvent.kt 定义了一个消息实体：

```
1. // 消息实体，具有两个属性 orderId 和 paidMoney
2. class OrderPaidEvent(
3.     val orderId: String,
```

```kotlin
4.         val paidMoney: BigDecimal
5.  ){
6.      constructor():this("0", BigDecimal(0.0))
7.      override fun toString(): String {
8.          return "OrderPaidEvent($orderId, $paidMoney)"
9.      }
10. }
```

MqProducer.kt 定义消息的生产者:

```kotlin
1.  @Component
2.  class MqProducer {
3.      // 注入 rocketMQTemplate
4.      @Autowired
5.      lateinit var rocketMQTemplate: RocketMQTemplate
6.      // 向"kotlin-topic"发送一条消息
7.      fun sendMessage(orderPaidEvent: OrderPaidEvent) {
8.          println("send message: $orderPaidEvent")
9.          rocketMQTemplate.send("kotlin-topic", MessageBuilder.withPayload
    (orderPaidEvent).build())
10.     }
11.     // 向"kotlin-topic"发送一条消息，tag 是 kotlin-tag
12.     fun sendMessageWithTag(orderPaidEvent: OrderPaidEvent) {
13.         println("send message: $orderPaidEvent, tag: kotlin-tag")
14.         rocketMQTemplate.convertAndSend("kotlin-topic:kotlin-tag", orderPaidEvent)
15.     }
16.     // 使用 convertAndSend 方法向"kotlin-topic"发送一条消息
17.     fun convertAndSendMessage(orderPaidEvent: OrderPaidEvent) {
18.         println("convertAndSend message: $orderPaidEvent")
19.         rocketMQTemplate.convertAndSend("kotlin-topic", orderPaidEvent)
20.     }
21.     // 向"kotlin-topic"异步发送一条消息，发送成功后执行 SendCallback 回调函数
22.     fun asyncSendMessage(orderPaidEvent: OrderPaidEvent) {
23.         println("async send single message: $orderPaidEvent")
24.         rocketMQTemplate.asyncSend("kotlin-topic", orderPaidEvent, object:
    SendCallback{
25.             override fun onSuccess(p0: SendResult?) {
```

```kotlin
26.            println("async send success: $p0")
27.        }
28.        override fun onException(p0: Throwable?) {
29.            throw Exception(p0)
30.        }
31.    },1000L)
32. }
33. // 向"kotlin-topic"批量同步发送多条消息，超时时间为 60 秒
34. fun syncSendBatchMessage(orderPaidEvents: List<OrderPaidEvent>) {
35.     println("sync send single message: $orderPaidEvents")
36.     val msg = orderPaidEvents.map { o -> MessageBuilder.withPayload(o).build() }
37.     rocketMQTemplate.syncSend("kotlin-topic", msg, 60000L)
38. }
39. // 向"kotlin-topic"同步发送一条消息
40. fun syncSendMessage(orderPaidEvent: OrderPaidEvent) {
41.     println("sync send single message: $orderPaidEvent")
42.     rocketMQTemplate.syncSend("kotlin-topic", orderPaidEvent)
43. }
44. }
```

MqConsumer.kt 定义消息的消费者：

```kotlin
1. @Component
2. //开启@RocketMQ 消息监听注解，消费 kotlin-topic 的消息，消费组是 kotlin-consumer
3. @RocketMQMessageListener(topic = "kotlin-topic", consumerGroup = "kotlin-consumer")
4. class MqConsumer: RocketMQListener<OrderPaidEvent> {
5.     // 消费消息后打印
6.     override fun onMessage(p0: OrderPaidEvent?) {
7.         println("OrderPaidEventConsumer received: $p0")
8.     }
9. }
```

MqTagConsumer.kt 定义消费的消息的 tag 是 "kotlin-tag"：

```kotlin
1. @Component
2. // 开启@RocketMQ 消息监听注解，消费 kotlin-topic 的消息，消费组是 kotlin-consumer1
3. // 只消费 tag 是 kotlin-tag 的消息
4. @RocketMQMessageListener(topic = "kotlin-topic", consumerGroup =
```

```kotlin
        "kotlin-consumer1", selectorExpression = "kotlin-tag")
5.  class MqTagConsumer: RocketMQListener<OrderPaidEvent> {
6.      // 消费消息后打印
7.      override fun onMessage(p0: OrderPaidEvent?) {
8.          println("OrderPaidEventConsumer received: $p0")
9.      }
10. }
```

MqController.kt 定义了几个接口测试 RocketMQ：

```kotlin
1.  @RestController
2.  class MqController {
3.      @Autowired
4.      lateinit var mqProducer: MqProducer
5.      // 测试接口，测试发送单条消息
6.      @PostMapping("/mq/send")
7.      fun sendMsg(@RequestBody orderPaidEvent: OrderPaidEvent) {
8.          mqProducer.sendMessage(orderPaidEvent)
9.      }
10.     // 测试接口，测试发送单条消息，并打标签
11.     @PostMapping("/mq/send/tag")
12.     fun sendMsgTag(@RequestBody orderPaidEvent: OrderPaidEvent) {
13.         mqProducer.sendMessageWithTag(orderPaidEvent)
14.     }
15.     // 测试接口，测试使用 convertAndSendMessage 方法发送单条消息
16.     @PostMapping("/mq/convertAndSend")
17.     fun convertAndSendMsg(@RequestBody orderPaidEvent: OrderPaidEvent) {
18.         mqProducer.convertAndSendMessage(orderPaidEvent)
19.     }
20.     // 测试接口，测试异步发送单条消息
21.     @PostMapping("/mq/asyncSend")
22.     fun asyncAndSendMsg(@RequestBody orderPaidEvent: OrderPaidEvent) {
23.         mqProducer.asyncSendMessage(orderPaidEvent)
24.     }
25.     // 测试接口，测试批量同步发送消息
26.     @PostMapping("/mq/asyncBatchSend")
27.     fun asyncAndBatchSendMsg(@RequestBody orderPaidEvents: List<OrderPaidEvent>) {
```

```
28.        mqProducer.syncSendBatchMessage(orderPaidEvents)
29.    }
30.    // 测试接口，测试同步发送单条消息
31.    @PostMapping("/mq/syncSend")
32.    fun syncSendMsg(@RequestBody orderPaidEvent: OrderPaidEvent) {
33.        mqProducer.syncSendMessage(orderPaidEvent)
34.    }
35. }
```

sendMsg 方法向 topic："kotlin-topic" 发送一条消息 orderPaidEvent，MqConsumer 会消费这条消息。sendMsgTag 方法发送一条带有 tag 标签的消息，tag 是 "kotlin-tag"，MqConsumer 和 MqTagConsumer 都会消费这条消息，但是 MqTagConsumer 只会消费 tag 是 "kotlin-tag" 的消息。convertAndSendMsg 方法用 convertAndSend 方法发送了一条消息，convertAndSend 方法会对消息实体进行包装，然后发送。asyncAndSendMsg 异步发送一条消息，并定义了回调方法 SendCallBack，消息发送成功会执行该方法。asyncAndBatchSendMsg 异步批量发送消息，如图 4.5 所示，批量发送三条消息。syncSendMsg 同步发送一条消息。

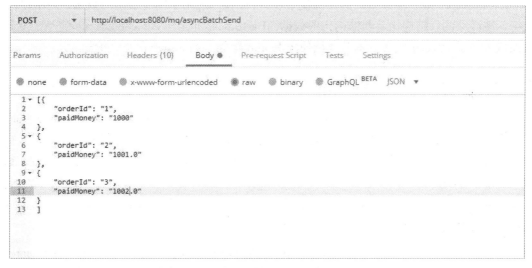

图4.5　消息队列批量发送消息接口参数

使用 RocketMQ-Console-Ng 可以查看 "kotlin-topic" 的消费情况。kotlin-topic 有 4 个队列，其中队列 0、2、3 各发送了 2、2、1 条消息，brokerOffset 和 consumerOffset 相等，没有出现消费滞后，如图 4.6 所示。

Topic	kotlin-topic	Delay	0	LastConsumeTime	2020-01-29 10:59:06

broker	queue	consumerClient	brokerOffset	consumerOffset	diffTotal	lastTimestamp
DESKTOP-644KAN2	0		2	2	0	2020-01-29 10:59:06
DESKTOP-644KAN2	1		0	0	0	1970-01-01 08:00:00
DESKTOP-644KAN2	2		2	2	0	2020-01-29 10:58:46
DESKTOP-644KAN2	3		1	1	0	2020-01-29 10:58:56

图4.6　kotlin-topic消费情况

4.7　Kotlin 集成 Elasticsearch

Elasticsearch 是一个分布式的开源搜索和分析引擎，适用于所有类型的数据，包括文本、数字、地理空间、结构化及非结构化的数据。Elasticsearch 是在 Apache Lucene 的基础上开发而成的，于 2010 年发布。本节介绍使用 Kotlin 集成 Elasticsearch 进行开发。

4.7.1　Elasticsearch 介绍

Elasticsearch 具有简单的 REST 风格 API，兼具分布式特性和高可扩展性，是 Elastic Stack 的核心组件。Elastic Stack 是适用于数据采集、存储、分析、可视化的一组开源工具，包括 Elasticsearch、Logstash、Kibana 等。Elastic Stack 简化了数据采集、可视化和报告过程，通过与 Beats 和 Logstash 进行集成，用户能够在 Elasticsearch 中索引数据之前轻松地处理数据。Kibana 不仅可针对 Elasticsearch 数据提供实时可视化，同时还提供 UI 以便用户快速访问应用程序性能监测（APM）、日志和基础设施指标等数据。

原始数据会从多个来源（包括日志、系统指标和网络应用程序）被输入到 Elasticsearch 中。数据采集是在 Elasticsearch 中进行索引之前解析、标准化并充实这些原始数据的过程。这些数据在 Elasticsearch 中索引完成之后，用户便可针对数据进行复杂的查询，并使用聚合来检索自身数据的复杂汇总。

Elasticsearch 索引指相互关联的文档集合。Elasticsearch 会以 JSON 文档的形式存储数据。每个文档都会在一组键（字段或属性的名称）和它们对应的值（字符串、数字、布尔值、日期、数值组、地理位置或其他类型的数据）之间建立联系。

Elasticsearch 使用的是一种被称为倒排索引的数据结构，这种结构允许快速地进行全文本搜索。倒排索引会列出在所有文档中出现的每个特有词汇，并且可以找到包含每个词汇的全部文档。

在索引过程中，Elasticsearch 会存储文档并构建倒排索引，这样用户便可以近实时地对文档数据进行搜索。索引过程是在索引 API 中启动的，通过此 API，你既可向特定索引中添加 JSON 文档，也可更改特定索引中的 JSON 文档。

Elasticsearch 的检索速度很快。由于 Elasticsearch 是在 Lucene 基础上构建而成的，所以在全文本搜索方面表现十分出色。Elasticsearch 同时还是一个近实时的搜索平台，这意味着从文档索引操作到文档变为可搜索状态之间的延迟很短，一般只有一秒。因此，Elasticsearch 非常适用于对时间有严苛要求的用例，例如，安全分析和基础设施监测。

Elasticsearch 具有分布式的本质特征。Elasticsearch 中存储的文档分布在不同的容器中，这些容器被称为分片，可以进行复制以提供数据冗余副本，以防发生硬件故障。Elasticsearch 的分布式特征使得它可以扩展至数百台（甚至数千台）服务器，并处理 PB 量级的数据。

Elasticsearch 还包含一系列广泛的功能。除了速度、可扩展性和弹性等优势以外，它还有大量强大的内置功能（例如，数据汇总和索引生命周期管理），可以方便用户更加高效地存储和搜索数据。

Elasticsearch 支持多种编程语言，官方提供了针对 Java、JavaScript（Node.js）、Go、.NET（C#）、PHP、Perl、Python、Ruby 语言的客户端。Elasticsearch 提供强大且全面的 REST API 集合，这些 API 可用来执行各种任务，例如，检查集群的运行状况、针对索引执行 CRUD（创建、读取、更新、删除）和搜索操作，以及执行诸如筛选和聚合等高级搜索操作。

Elasticsearch 能够索引多种类型的内容，可应用于应用程序搜索、网站搜索、企业搜索、日志处理和分析、基础设施指标和容器监测、应用程序性能监测、地理空间数据分析、可视化、安全分析及业务分析等场景。

4.7.2 使用 Kotlin 操作 Elasticsearch

Kotlin 集成 Elasticsearch 需要在 pom.xml 中添加如下依赖：

```
1.  <!-- Spring Boot Elasticsearch 依赖包 -->
2.  <dependency>
3.      <groupId>org.springframework.boot</groupId>
```

```xml
4.     <artifactId>spring-boot-starter-data-elasticsearch</artifactId>
5. </dependency>
```

在 application.yml 中添加如下配置，配置 Elasticsearch 的集群，我们使用的是单机 Elasticsearch：

```yaml
1. spring:
2.   data:
3.     elasticsearch:
4.       cluster-name: elasticsearch      #集群名称
5.       cluster-nodes: 127.0.0.1:9300    #elasticsearch host port
```

Item.kt 定义了 Elasticsearch 中的一个实体对象：

```kotlin
1. // 定义实体 Item
2. @Document(indexName = "item",type = "docs", shards = 1, replicas = 0)
3. data class Item(
4.         @Id
5.         val id: Long,
6.         // 对 title 使用 ik_max_word 进行分词
7.         @Field(type = FieldType.Text, analyzer = "ik_max_word")
8.         val title: String,
9.         @Field(type = FieldType.Keyword)
10.        val category: String,
11.        @Field(type = FieldType.Keyword)
12.        val brand: String,
13.        @Field(type = FieldType.Double)
14.        val price: Double,
15.        @Field(index = false, type = FieldType.Keyword)
16.        val images: String
17. ){
18.     constructor():
19.             this(0L, "", "", "", 0.0, "")
20. }
```

ItemRepository.kt 定义了 Item 集合的 CRUD 操作：

```kotlin
1. interface ItemRepository: ElasticsearchRepository<Item, Long>
```

ItemRepositoryTest.kt 对 Item 集合的常见操作进行了测试。createIndex、putMapping 方法分别创建了 Item 索引和映射：

```kotlin
1.  @SpringBootTest
2.  @TestMethodOrder(MethodOrderer.OrderAnnotation::class)
3.  @ExtendWith(SpringExtension::class)
4.  class ItemRepositoryTest {
5.      @Autowired
6.      lateinit var elasticsearchTemplate: ElasticsearchTemplate
7.
8.      @Autowired
9.      lateinit var itemRepository: ItemRepository
10.     // 创建索引、映射
11.     @Test
12.     @Order(1)
13.     fun testCreateIndex() {
14.         elasticsearchTemplate.createIndex(Item::class.java)
15.         elasticsearchTemplate.putMapping(Item::class.java)
16.     }
17. }
```

保存一条新的 Item 记录：

```kotlin
1.  // 测试插入一条 Item 记录
2.  @Test
3.  @Order(2)
4.  fun testAddNewItem() {
5.      val item = Item(1L, "Iphone 11", "手机", "苹果", 5899.0,
    "http://image.baidu.com/13123.jpg")
6.      itemRepository.save(item)
7.  }
```

保存多条 Item 记录：

```kotlin
1.  // 测试批量插入 Item 记录
2.  @Test
3.  @Order(3)
4.  fun testBatchAddNewItems(){
```

```
5.      val items = arrayOf(
6.          Item(2L, "坚果手机 R1", " 手机", "锤子", 3699.00,
   "http://image.baidu.com/13123.jpg"),
7.          Item(3L, "华为 MATE10", " 手机", "华为", 4499.00,
   "http://image.baidu.com/13123.jpg")
8.      )
9.      itemRepository.saveAll(items.asList())
10. }
11.
```

更新 id=1 的 Item 记录:

```
1. // 测试更新 Item 记录
2. @Test
3. @Order(4)
4. fun testUpdateItem(){
5.     val item = Item(1L, "苹果XS Max", " 手机", "苹果", 4899.00,
   "http://image.baidu.com/13123.jpg")
6.     itemRepository.save(item)
7. }
```

查找所有的 Item 记录:

```
1. // 测试查找 Item 集合中的所有记录
2. @Test
3. @Order(5)
4. fun testFindAll(){
5.     val items = itemRepository.findAll()
6.     Assert.assertEquals(3, items.toList().size)
7. }
```

分页查找 Item 记录,每页返回 2 条记录,查找第 1 页的记录:

```
1. // 测试分页查找 Item 记录
2. @Test
3. @Order(6)
4. fun testFindByPage(){
5.     val items = itemRepository.findAll(PageRequest.of(1, 2))
6.     items.forEach { println(it) }
```

7. }

查找所有 Item 记录，并根据价格降序排列：

```
1.  // 测试根据 price 降序查找 Item 记录
2.  @Test
3.  @Order(7)
4.  fun testFindBySort(){
5.      val items = itemRepository.findAll(Sort.by("price").descending())
6.      items.forEach { println(it) }
7.  }
```

查找 title 是"坚果手机"的 Item 记录，matchQuery 进行词条匹配，先分词然后再查询结果：

```
1.  // 测试查找 title 包含"坚果手机"的 Item 记录，进行分词
2.  @Test
3.  @Order(8)
4.  fun testMatchQuery(){
5.      val queryBuilder = NativeSearchQueryBuilder()
6.      queryBuilder.withQuery(QueryBuilders.matchQuery("title", "坚果手机"))
7.      val items = itemRepository.search(queryBuilder.build())
8.      Assert.assertEquals(1, items.totalElements)
9.      items.forEach { println(it) }
10. }
```

查找 title 是"坚果"的 Item 记录，使用 termQuery 进行词条匹配，不分词：

```
1.  // 测试查找 title 包含"坚果"的 Item 记录，不分词
2.  @Test
3.  @Order(9)
4.  fun testTermQuery(){
5.      val queryBuilder = NativeSearchQueryBuilder()
6.      queryBuilder.withQuery(QueryBuilders.termQuery("title", "坚果"))
7.      val items = itemRepository.search(queryBuilder.build())
8.      Assert.assertEquals(1, items.totalElements)
9.      items.forEach { println(it) }
10. }
```

查找 title 包含"坚果"的 Item 记录，使用 fuzzyQuery 进行模糊查询：

```
1.  // 测试查找 title 包含"坚果"的 Item 记录, 模糊查询
2.  @Test
3.  @Order(10)
4.  fun testFuzzyQuery(){
5.      val queryBuilder = NativeSearchQueryBuilder()
6.      queryBuilder.withQuery(QueryBuilders.fuzzyQuery("title", "坚果"))
7.      val items = itemRepository.search(queryBuilder.build())
8.      Assert.assertEquals(1, items.totalElements)
9.      items.forEach { println(it) }
10. }
```

查找 title 是"坚果"、brand 是"锤子"的 Item 记录。使用 booleanQuery，布尔查询可查询布尔关系，包括 BooleanClause.Occur.MUST、BooleanClause.Occur.MUST_NOT、BooleanClause.Occur.SHOULD，分别表示必须包含、不能包含和可以包含三种。

```
1.  // 测试查找 title 包含"坚果", brand 是"锤子"的 Item 记录
2.  @Test
3.  @Order(11)
4.  fun testBooleanQuery(){
5.      val queryBuilder = NativeSearchQueryBuilder()
6.      queryBuilder.withQuery(QueryBuilders.boolQuery()
7.              .must(QueryBuilders.termQuery("title", "坚果"))
8.              .must(QueryBuilders.termQuery("brand", "锤子")))
9.
10.     val items = itemRepository.search(queryBuilder.build())
11.     Assert.assertEquals(1, items.totalElements)
12.     items.forEach { println(it) }
13. }
```

查找价格在 3000 元到 4000 元的 Item 记录，使用 rangeQuery 可进行范围查找：

```
1.  // 测试查找价格在 3000 元到 4000 元之间的 Item 记录
2.  @Test
3.  @Order(12)
4.  fun testRangeQuery(){
5.      val queryBuilder = NativeSearchQueryBuilder()
```

```
6.     queryBuilder.withQuery(QueryBuilders.rangeQuery("price").from(3000).to(4000))
7.
8.     val items = itemRepository.search(queryBuilder.build())
9.     Assert.assertEquals(2, items.totalElements)
10.    items.forEach { println(it) }
11. }
```

删除 id=1 的 Item 记录:

```
1. // 测试删除 Item 记录
2. @Test
3. @Order(13)
4. fun testDelete(){
5.     itemRepository.deleteById(1)
6.
7.     Assert.assertEquals(2, itemRepository.count())
8. }
```

4.8 Kotlin 集成 Swagger

Swagger 是一个规范和完整的框架,用于生成、描述、调用和可视化 RESTful 风格的 Web 服务。本节介绍使用 Kotlin 集成 Swagger 开发。

4.8.1 Swagger 介绍

Swagger 提供了构建 API 的一套工具和规范,按照它的规范去定义接口及接口相关的信息,再通过 Swagger 衍生出来的一系列项目和工具,就可以生成各种格式的接口文档,生成多种语言的客户端和服务端的代码,以及生成接口调试页面等。在开发新版本或者迭代版本的时候,只需更新 Swagger 描述文件,就可以自动生成接口文档、客户端及服务端代码,做到调用端代码、服务端代码以及接口文档的一致性。

Swagger 包括如下开源项目。

- **Swagger Codegen**:通过 Codegen 可以将描述文件生成 HTML 格式或 cwiki 形式的接口文档,同时也能生成多种语言的服务端和客户端的代码。支持通过 Jar 包、Docker、Node.js 等方式在本地化运行生成,也可以在后面介绍的 Swagger Editor

中在线生成。
- **Swagger UI**：提供了一个可视化的 UI 页面展示描述文件。接口的调用方、测试方、项目经理等都可以在该页面中对相关接口进行查阅和做一些简单的接口请求。该项目支持在线导入描述文件和在本地部署 UI 项目。
- **Swagger Editor**：编辑 Swagger 描述文件的编辑器，该编辑器支持实时预览描述文件的更新效果，还提供了在线编辑器和本地部署编辑器两种方式。
- **Swagger Inspector**：可以对接口进行在线测试，比在 Swagger UI 中做接口请求，会返回更多的信息，也会保存用户的实际请求参数等数据。
- **Swagger Hub**：集成了上面所有项目的各项功能，以项目和版本为单位，将用户的描述文件上传到 Swagger Hub 中。在 Swagger Hub 中可以完成上面介绍的项目的所有工作，需要注册账号，有免费版和收费版。

维护这个 JSON 或 YAML 格式的描述文件有一定的工作量，在持续迭代开发的时候，往往会忽略更新这个描述文件，导致基于该描述文件生成的接口文档失去了参考意义。Springfox Swagger 基于 Swagger 规范，可以自动检查类、控制器、方法、模型类以及它们映射到的 URL，自动生成 JSON 格式的描述文件，进而生成与代码一致的接口文档和客户端代码。

Swagger 常用的注解有@Api()，用于类，表示这个类是 Swagger 的资源；@ApiOperation()，用于方法，表示一个 HTTP 请求的操作；@ApiParam()，用于方法、参数和字段说明，表示对参数添加元数据（说明或是否必填等）；@ApiModel()，用于类，对类进行说明；@ApiModelProperty()，用于方法、字段，对属性字段进行说明或者对数据操作更改进行说明；@ApiIgnore()，用于类、方法、方法参数，表示这个方法或者类被忽略；@ApiImplicitParam()，用于方法，表示这是单独的请求参数；@ApiImplicitParams()用于方法，包含多个@ApiImplicitParam。

对于前后端分离开发的场景，如果能在提供接口文档的同时，把所有接口的模拟请求响应数据也提供给前端，或者有 Mock 系统，直接将这些模拟数据录入 Mock 系统，那将会提高前端的开发效率，减少许多发生在联调时才会发生的问题。通过适当地在代码中加入 Swagger 的注解，可以让接口文档描述信息更加详细，如果把每个出入参数的示例值都配上，那前端就可以直接在接口文档中拿到模拟数据。

4.8.2 使用 Kotlin 操作 Swagger

Kotlin 集成 Swagger 需要在 pom.xml 中添加如下依赖：

```xml
1. <-- Swagger 依赖包 -->
2. <dependency>
3.     <groupId>io.springfox</groupId>
4.     <artifactId>springfox-swagger2</artifactId>
5.     <version>2.9.2</version>
6. </dependency>
7. <dependency>
8.     <groupId>io.springfox</groupId>
9.     <artifactId>springfox-swagger-ui</artifactId>
10.    <version>2.9.2</version>
11. </dependency>
```

SwaggerConfig.kt 定义了接口的基本信息：

```kotlin
1. @Configuration
2. @EnableSwagger2
3. class SwaggerConfig {
4.     // 配置 apiInfo，定义包路径，扫描该路径下的所有接口
5.     @Bean
6.     fun createRestApi(): Docket {
7.         return Docket(DocumentationType.SWAGGER_2)
8.             .apiInfo(apiInfo())
9.             .select()
10.            .apis(RequestHandlerSelectors.basePackage("io.kang.example.controller.swagger"))
11.            .paths(PathSelectors.any())
12.            .build()
13.    }
14.    // 配置 apiInfo 的具体信息：标题、联系人、版本号、描述
15.    @Bean
16.    fun apiInfo(): ApiInfo {
17.        return ApiInfoBuilder()
18.            .title("使用 Swagger2 构建 RESTful APIs")
19.            .contact(Contact("dutyk", "https://github.com/dutyk", "1013812851@qq.com"))
```

```
20.            .version("1.0")
21.            .description("Demo for book Kotlin Spring Boot Action")
22.            .build()
23.    }
24. }
```

SwaggerController.kt 定义了 get、post、put、delete 接口，分别对应 GetMapping、PostMapping、PutMapping 及 DeleteMapping，并用@ApiOperation()添加了对接口的描述，用@ApiImplicitParam()添加了对参数的描述。

```
1.  @RestController
2.  @RequestMapping("/users")
3.  class SwaggerController {
4.      @Autowired
5.      lateinit var userRepository: UserRepository
6.      // get 接口，获取用户列表
7.      @ApiOperation(value="获取用户列表", notes="")
8.      @GetMapping("/all")
9.      fun getUserList(): List<User> {
10.         return userRepository.findAll().toList()
11.     }
12.     // post 接口，新增用户
13.     @ApiOperation(value="创建用户", notes="根据 User 对象创建用户")
14.     @ApiImplicitParam(name = "user", value = "用户详细实体 user", required = true,
    dataType = "User")
15.     @PostMapping("/add")
16.     fun postUser(@RequestBody user: User): String {
17.         userRepository.save(user)
18.         return "success"
19.     }
20.     // get 接口，在接口传递参数：id，获取指定用户信息
21.     @ApiOperation(value="获取用户详细信息", notes="根据 url 的 id 来获取用户详细信息")
22.     @ApiImplicitParam(name = "id", value = "用户 ID", required = true, dataType =
    "Long")
23.     @GetMapping("/find/{id}")
24.     fun getUser(@PathVariable id: Long): User? {
25.         return userRepository.findById(id).get()
```

```kotlin
26.    }
27.    // put 接口，更新指定用户的信息
28.    @ApiOperation(value="更新用户详细信息", notes="根据 url 的 id 来指定更新对象，并根据传过来的 user 信息来更新用户详细信息")
29.    @ApiImplicitParams(
30.        ApiImplicitParam(name = "id", value = "用户 ID", required = true, dataType = "Long"),
31.        ApiImplicitParam(name = "user", value = "用户详细实体 user", required = true, dataType = "User")
32.    )
33.    @PutMapping("/update/{id}")
34.    fun updateUser(@PathVariable id: Long, @RequestBody user: User) {
35.        val user = User(id, user.userName, user.password, user.email, user.age, user.height, user.address, user.education, user.income)
36.        userRepository.save(user)
37.    }
38.    // delete 接口，删除指定用户
39.    @ApiOperation(value="删除用户", notes="根据 url 的 id 来指定删除对象")
40.    @ApiImplicitParam(name = "id", value = "用户 ID", required = true, dataType = "Long")
41.    @DeleteMapping("/delete/{id}")
42.    fun deleteUser(@PathVariable id: Long) {
43.        userRepository.deleteById(id)
44.    }
45. }
```

在浏览器中输入 swagger-ui 的地址，可跳转到接口调试页面对定义的接口进行测试。

"获取用户列表"接口测试如下：单击"Execute"可以测试这个接口，返回用户列表，这个测试的是 get 请求，如图 4.7 所示。

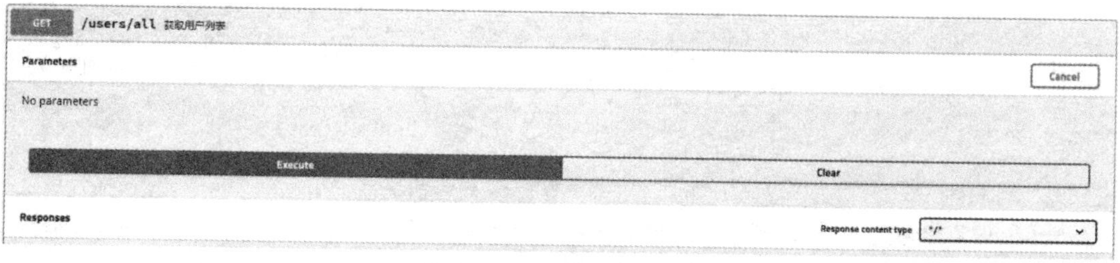

图4.7　获取用户列表接口

接口的返回结果如图4.8所示。

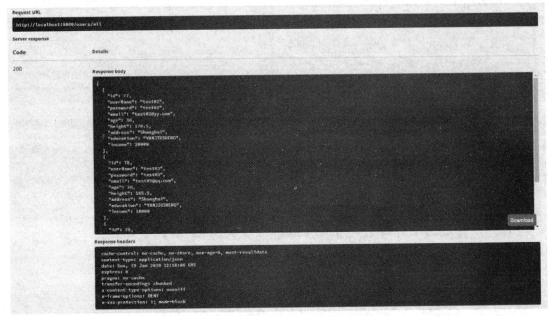

图4.8 "获取用户列表"接口返回的结果

"创建用户"接口测试如下：输入用户的各个属性的值，提交即可。这个测试的是post请求，如图4.9所示。

图4.9 "创建用户"接口

"根据 id 查找用户"接口测试如图 4.10 所示，在输入框中输入用户 id，这个测试的是 get 请求。

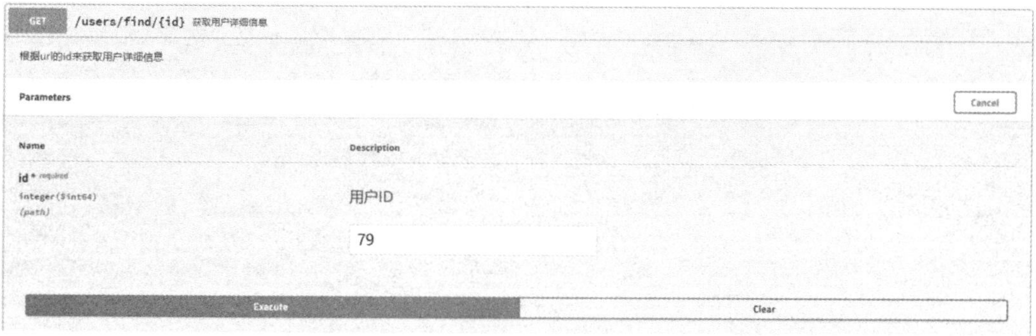

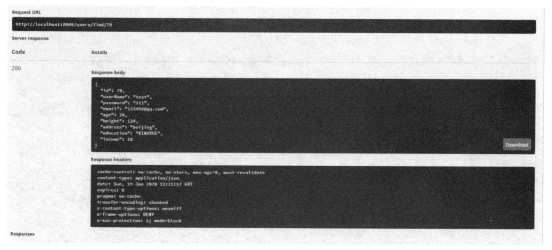

图4.10 "根据id查找用户"接口

接口返回 id 是 79 的用户记录，如图 4.11 所示。

图4.11 "根据id查找用户"接口返回的记录

"更新指定 id 用户信息"接口测试如下：在输入框中输入用户的 id 为 79，并填写用户的信息。这个测试的是 put 请求，如图 4.12 所示。

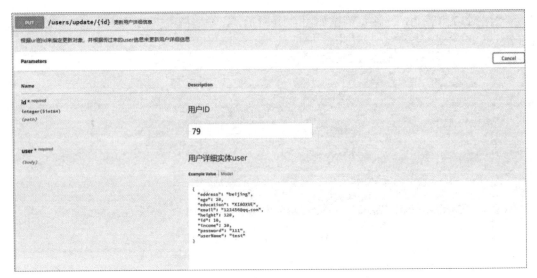

图4.12 "更新指定id用户信息"接口

"根据 id 删除用户"接口测试如下：输入用户 id=79，可以把 id=79 的记录删除，此时，查不到 id=79 的用户记录 3。这个测试的是 delete 请求，如图 4.13 所示。

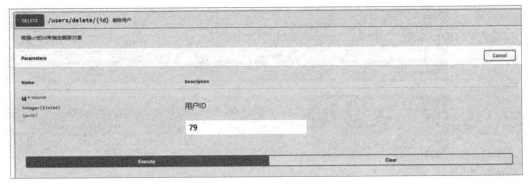

图4.13 "根据id删除用户"接口

4.9 小结

本章介绍了使用 Kotlin 操作常用的微服务组件的方法，包括的组件有：Spring Boot、Redis、JPA、QueryDSL、MongoDB、Spring Security、RocketMQ、Elasticsearch、Swagger。Kotlin 可以使用已有的 Jar 包，节省了开发成本。本章对每个组件都提供了若干示例，方便大家在使用时参考。

第 5 章
Kotlin 应用于微服务注册中心

微服务注册中心对于微服务系统很重要,有助于系统解耦,开发分布式应用系统。本章介绍将 Kotlin 应用于微服务注册中心的相关知识,主要包括四种常用的注册中心:Eureka、Consul、Zookeeper、Nacos。

5.1 Eureka

Eureka 是 Netflix 公司开发的服务发现框架,是一个基于 REST 的服务。它主要以 AWS 云服务为支撑,提供服务发现并实现负载均衡和故障转移。Spring Cloud 将它集成在其子项目 spring-cloud-netflix 中,以实现 Spring Cloud 的服务发现功能。本节介绍使用 Eureka 作为 Kotlin 开发的微服务注册中心的相关知识。

5.1.1 Eureka 介绍

Eureka 有 Server 和 Client 两个组件。服务端提供服务注册,当客户端服务启动的时候,会主动向服务端进行注册,服务端会存储所有已经注册的服务节点的信息。服务端会管理这些节点信息,并且会将异常的节点从服务列表中移除。客户端有缓存功能,所以即便 Eureka 集群中的所有节点都失效,或者发生网络分区故障导致客户端不能访问任何一台 Eureka 服务器,Eureka 服务的消费者仍然可以通过 Eureka 客户端缓存来获取现有的服务注册信息。无论是服务端还是客户端，都支持集群模式，注册信息和更新信息会在整个 Eureka 集群的节点中进行复制。

  Eureka Client 是一个 Java 客户端，用于简化与 Eureka Server 的交互，客户端同时具备一个内置的、使用轮询（round-robin）负载算法的负载均衡器。在应用启动后，应用将会向 Eureka Server 发送心跳，默认周期为 30 秒，如果 Eureka Server 在多个心跳周期内没有接收到某个节点的心跳，Eureka Server 会从服务注册列表中把这个服务节点移除（默认 90 秒）。Eureka Client 分为两个角色：Application Service（Service Provider）和 Application Client（Service Consumer）。服务提供方是注册到 Eureka Server 中的服务，服务消费方是通过 Eureka Server 发现服务，并消费。Eureka Server 的架构如图 5.1 所示。

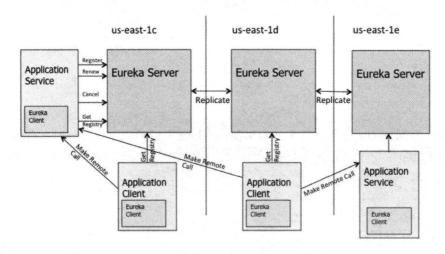

图5.1 Eureka Server的架构图

  Register（服务注册）：把自己的 IP 地址和端口注册给 Eureka。

  Renew（服务续约）：发送心跳包，每 30 秒发送一次，告诉 Eureka 自己还活着。

  Cancel（服务下线）：当 Provider 关闭时会向 Eureka 发送消息，把自己从服务列表中删除，防止 Consumer 调用到不存在的服务。

  Get Registry（获取服务注册列表）：获取其他服务列表。

  Replicate（集群中数据同步）：Eureka 集群中的数据复制与同步。

  Make Remote Call（远程调用）：完成服务的远程调用。

5.1.2 Kotlin 集成 Eureka 服务注册

新建一个 Maven 子工程 chapter05-eureka，这个一是 Eureka Server，其 pom.xml 文件主要定义了 Eureka Server 相关的包：

```xml
1.  <?xml version="1.0" encoding="UTF-8"?>
2.  <project xmlns="http://maven.apache.org/POM/4.0.0"
3.          xmlns:xsi="http://www.w3.org/2001/XMLSchema-instance"
4.          xsi:schemaLocation="http://maven.apache.org/POM/4.0.0
    http://maven.apache.org/xsd/maven-4.0.0.xsd">
5.      <!-- 父 pom -->
6.      <parent>
7.          <artifactId>kotlinspringboot</artifactId>
8.          <groupId>io.kang.kotlinspringboot</groupId>
9.          <version>0.0.1-SNAPSHOT</version>
10.     </parent>
11.     <modelVersion>4.0.0</modelVersion>
12.     <!-- 子工程名称 -->
13.     <artifactId>chapter05-eureka</artifactId>
14. 
15.     <dependencies>
16.         <!-- Eureka 服务端依赖包 -->
17.         <dependency>
18.             <groupId>org.springframework.cloud</groupId>
19.             <artifactId>spring-cloud-starter-netflix-eureka-server</artifactId>
20.             <version>2.2.1.RELEASE</version>
21.         </dependency>
22.         <!-- Kotlin 相关依赖包 -->
23.         <dependency>
24.             <groupId>com.fasterxml.jackson.module</groupId>
25.             <artifactId>jackson-module-kotlin</artifactId>
26.         </dependency>
27.         <dependency>
28.             <groupId>org.jetbrains.kotlin</groupId>
29.             <artifactId>kotlin-reflect</artifactId>
30.         </dependency>
31.         <dependency>
```

```xml
32.            <groupId>org.jetbrains.kotlin</groupId>
33.            <artifactId>kotlin-stdlib-jdk8</artifactId>
34.        </dependency>
35.        <dependency>
36.            <groupId>org.jetbrains.kotlinx</groupId>
37.            <artifactId>kotlinx-coroutines-core</artifactId>
38.            <version>1.3.2</version>
39.        </dependency>
40.    </dependencies>
41.
42.    <build>
43.        <sourceDirectory>${project.basedir}/src/main/kotlin</sourceDirectory>
44.        <testSourceDirectory>${project.basedir}/src/test/kotlin</testSourceDirectory>
45.        <plugins>
46.            <plugin>
47.                <groupId>org.springframework.boot</groupId>
48.                <artifactId>spring-boot-maven-plugin</artifactId>
49.            </plugin>
50.            <plugin>
51.                <groupId>org.jetbrains.kotlin</groupId>
52.                <artifactId>kotlin-maven-plugin</artifactId>
53.                <configuration>
54.                    <args>
55.                        <arg>-Xjsr305=strict</arg>
56.                    </args>
57.                    <compilerPlugins>
58.                        <plugin>spring</plugin>
59.                        <plugin>jpa</plugin>
60.                    </compilerPlugins>
61.                </configuration>
62.                <dependencies>
63.                    <dependency>
64.                        <groupId>org.jetbrains.kotlin</groupId>
65.                        <artifactId>kotlin-maven-allopen</artifactId>
66.                        <version>${kotlin.version}</version>
67.                    </dependency>
68.                    <dependency>
```

```
69.                    <groupId>org.jetbrains.kotlin</groupId>
70.                    <artifactId>kotlin-maven-noarg</artifactId>
71.                    <version>${kotlin.version}</version>
72.                </dependency>
73.            </dependencies>
74.        </plugin>
75.    </plugins>
76. </build>
77. </project>
```

application.yml 的定义如下：

```
1. server:
2.   port: 8761  # 应用端口号
3. eureka:
4.   instance:
5.     hostname: localhost  # Eureka 实例的主机
6.   client:
7.     register-with-eureka: false  # 应用不注册到 Eureka
8.     fetch-registry: false  # 不从 Eureka 获取其他服务的地址
9.     service-url:  # Eureka 服务地址
10.      defaultZone: http://${eureka.instance.hostname}:${server.port}/eureka/
```

EurekaServerApplication.kt 定义了启动类：

```
1. // 开启 Eureka Server 注解
2. @EnableEurekaServer
3. @SpringBootApplication
4. class EurekaServerApplication
5. // 启动类
6. fun main(args: Array<String>) {
7.     runApplication<EurekaServerApplication>(*args);
8. }
```

运行这个类，启动一个单机的 Eureka Server。

5.1.3 一个 Eureka 服务提供方

新建一个 Maven 子工程：chapter05-eureka-provider，这是一个 Eureka Client，定义一

个服务提供方。pom.xml 文件内容如下：

```xml
1.  <?xml version="1.0" encoding="UTF-8"?>
2.  <project xmlns="http://maven.apache.org/POM/4.0.0"
3.          xmlns:xsi="http://www.w3.org/2001/XMLSchema-instance"
4.          xsi:schemaLocation="http://maven.apache.org/POM/4.0.0
    http://maven.apache.org/xsd/maven-4.0.0.xsd">
5.      <!-- 父 pom -->
6.      <parent>
7.          <artifactId>kotlinspringboot</artifactId>
8.          <groupId>io.kang.kotlinspringboot</groupId>
9.          <version>0.0.1-SNAPSHOT</version>
10.     </parent>
11.     <modelVersion>4.0.0</modelVersion>
12.     <!-- 子工程名称 -->
13.     <artifactId>chapter05-eureka-provider</artifactId>
14.
15.     <dependencies>
16.         <!-- Eureka 客户端依赖包 -->
17.         <dependency>
18.             <groupId>org.springframework.cloud</groupId>
19.             <artifactId>spring-cloud-starter-netflix-eureka-client</artifactId>
20.             <version>2.2.1.RELEASE</version>
21.         </dependency>
22.         <!-- Spring Boot Web 依赖包 -->
23.         <dependency>
24.             <groupId>org.springframework.boot</groupId>
25.             <artifactId>spring-boot-starter-web</artifactId>
26.             <version>2.2.1.RELEASE</version>
27.         </dependency>
28.         <dependency>
29.             <groupId>com.fasterxml.jackson.module</groupId>
30.             <artifactId>jackson-module-kotlin</artifactId>
31.         </dependency>
32.         <dependency>
33.             <groupId>org.jetbrains.kotlin</groupId>
34.             <artifactId>kotlin-reflect</artifactId>
```

```xml
35.        </dependency>
36.        <dependency>
37.            <groupId>org.jetbrains.kotlin</groupId>
38.            <artifactId>kotlin-stdlib-jdk8</artifactId>
39.        </dependency>
40.        <dependency>
41.            <groupId>org.jetbrains.kotlinx</groupId>
42.            <artifactId>kotlinx-coroutines-core</artifactId>
43.            <version>1.3.2</version>
44.        </dependency>
45.    </dependencies>
46.
47.    <build>
48.        <sourceDirectory>${project.basedir}/src/main/kotlin</sourceDirectory>
49.        <testSourceDirectory>${project.basedir}/src/test/kotlin</testSourceDirectory>
50.        <plugins>
51.            <plugin>
52.                <groupId>org.springframework.boot</groupId>
53.                <artifactId>spring-boot-maven-plugin</artifactId>
54.            </plugin>
55.            <plugin>
56.                <groupId>org.jetbrains.kotlin</groupId>
57.                <artifactId>kotlin-maven-plugin</artifactId>
58.                <configuration>
59.                    <args>
60.                        <arg>-Xjsr305=strict</arg>
61.                    </args>
62.                    <compilerPlugins>
63.                        <plugin>spring</plugin>
64.                        <plugin>jpa</plugin>
65.                    </compilerPlugins>
66.                </configuration>
67.                <dependencies>
68.                    <dependency>
69.                        <groupId>org.jetbrains.kotlin</groupId>
70.                        <artifactId>kotlin-maven-allopen</artifactId>
71.                        <version>${kotlin.version}</version>
```

```xml
72.        </dependency>
73.        <dependency>
74.            <groupId>org.jetbrains.kotlin</groupId>
75.            <artifactId>kotlin-maven-noarg</artifactId>
76.            <version>${kotlin.version}</version>
77.        </dependency>
78.       </dependencies>
79.      </plugin>
80.    </plugins>
81.  </build>
82. </project>
```

application.yml 的定义如下：

```yaml
1. server:
2.   port: 8000    #应用端口号
3. spring:
4.   application:
5.     name: provider-server    #应用名称
6. eureka:
7.   client:
8.     service-url:    #Eureka Server 访问地址
9.       defaultZone: http://localhost:8761/eureka/
```

ProviderApplication.kt 定义了启动类，运行它可以启动一个 Eureka Client，并注册到 Eureka Server：

```kotlin
1. // 开启 Eureka Client 注解，注册到 Eureka
2. @EnableEurekaClient
3. @SpringBootApplication
4. class ProviderApplication
5. // 启动类
6. fun main(args: Array<String>) {
7.     runApplication<ProviderApplication>(*args)
8. }
```

ProviderController.kt 定义了一个接口，提供服务：

```
1.  @RestController
2.  class ProviderController {
3.      // 服务方测试接口
4.      @GetMapping("/provide")
5.      fun provide(): String {
6.          return "Hello From Provider"
7.      }
8.  }
```

5.1.4　Kotlin 集成 OpenFeign 服务调用

新建一个 Maven 子工程：chapter05-eureka-consumer，这是一个 Eureka Client，是一个服务消费者，调用 5.1.3 节定义的服务提供方。pom.xml 的内容如下：

```
1.  <?xml version="1.0" encoding="UTF-8"?>
2.  <project xmlns="http://maven.apache.org/POM/4.0.0"
3.      xmlns:xsi="http://www.w3.org/2001/XMLSchema-instance"
4.      xsi:schemaLocation="http://maven.apache.org/POM/4.0.0
    http://maven.apache.org/xsd/maven-4.0.0.xsd">
5.      <!-- 父 pom -->
6.      <parent>
7.          <artifactId>kotlinspringboot</artifactId>
8.          <groupId>io.kang.kotlinspringboot</groupId>
9.          <version>0.0.1-SNAPSHOT</version>
10.     </parent>
11.     <modelVersion>4.0.0</modelVersion>
12.     <!-- 子工程名称 -->
13.     <artifactId>chapter05-eureka-consumer</artifactId>
14.     <dependencies>
15.         <!-- Eureka Client 依赖包 -->
16.         <dependency>
17.             <groupId>org.springframework.cloud</groupId>
18.             <artifactId>spring-cloud-starter-netflix-eureka-client</artifactId>
19.             <version>2.2.1.RELEASE</version>
20.         </dependency>
21.         <!-- Spring Boot Web 依赖包 -->
22.         <dependency>
```

```xml
23.            <groupId>org.springframework.boot</groupId>
24.            <artifactId>spring-boot-starter-web</artifactId>
25.            <version>2.2.1.RELEASE</version>
26.        </dependency>
27.        <!--Spring Cloud OpenFeign 依赖包 -->
28.        <dependency>
29.            <groupId>org.springframework.cloud</groupId>
30.            <artifactId>spring-cloud-starter-openfeign</artifactId>
31.            <version>2.2.1.RELEASE</version>
32.        </dependency>
33.        <dependency>
34.            <groupId>com.fasterxml.jackson.module</groupId>
35.            <artifactId>jackson-module-kotlin</artifactId>
36.        </dependency>
37.        <dependency>
38.            <groupId>org.jetbrains.kotlin</groupId>
39.            <artifactId>kotlin-reflect</artifactId>
40.        </dependency>
41.        <dependency>
42.            <groupId>org.jetbrains.kotlin</groupId>
43.            <artifactId>kotlin-stdlib-jdk8</artifactId>
44.        </dependency>
45.        <dependency>
46.            <groupId>org.jetbrains.kotlinx</groupId>
47.            <artifactId>kotlinx-coroutines-core</artifactId>
48.            <version>1.3.2</version>
49.        </dependency>
50.    </dependencies>
51.
52.    <build>
53.        <sourceDirectory>${project.basedir}/src/main/kotlin</sourceDirectory>
54.        <testSourceDirectory>${project.basedir}/src/test/kotlin</testSourceDirectory>
55.        <plugins>
56.            <plugin>
57.                <groupId>org.springframework.boot</groupId>
58.                <artifactId>spring-boot-maven-plugin</artifactId>
59.            </plugin>
```

```xml
60.        <plugin>
61.            <groupId>org.jetbrains.kotlin</groupId>
62.            <artifactId>kotlin-maven-plugin</artifactId>
63.            <configuration>
64.                <args>
65.                    <arg>-Xjsr305=strict</arg>
66.                </args>
67.                <compilerPlugins>
68.                    <plugin>spring</plugin>
69.                    <plugin>jpa</plugin>
70.                </compilerPlugins>
71.            </configuration>
72.            <dependencies>
73.                <dependency>
74.                    <groupId>org.jetbrains.kotlin</groupId>
75.                    <artifactId>kotlin-maven-allopen</artifactId>
76.                    <version>${kotlin.version}</version>
77.                </dependency>
78.                <dependency>
79.                    <groupId>org.jetbrains.kotlin</groupId>
80.                    <artifactId>kotlin-maven-noarg</artifactId>
81.                    <version>${kotlin.version}</version>
82.                </dependency>
83.            </dependencies>
84.        </plugin>
85.    </plugins>
86. </build>
87. </project>
```

application.yml 的内容如下：

```yaml
1. server:
2.   port: 8001  #应用端口号
3. spring:
4.   application:
5.     name: consumer-feign  #应用名称
6. eureka:
```

```
7.   client:
8.     service-url:  #Eureka 服务访问地址
9.       defaultZone: http://localhost:8761/eureka/
```

ConsumerApplication.kt 定义了一个启动类，启动了一个消费者，采用 Feign 进行服务间调用，并注册到 Eureka Server：

```
1. // 开启 Feign 和 Eureka Client 注解
2. @EnableFeignClients
3. @EnableEurekaClient
4. @SpringBootApplication
5. class ConsumerApplication
6. // 启动类
7. fun main(args: Array<String>) {
8.     runApplication<ConsumerApplication>(*args)
9. }
```

ProviderService.kt 定义了一个接口，调用 chapter05-eureka-provider 定义的接口，@FeignClient 指定要调用的服务名：

```
1. // 通过 Feign 调用 provider-server 服务
2. @FeignClient(value = "provider-server")
3. interface ProviderService {
4.     @GetMapping("/provide")
5.     fun provide(): String
6. }
```

ConsumerController.kt 定义了一个接口，可以测试 Feign 调用：

```
1. @RestController
2. class ConsumerController {
3.     @Autowired
4.     lateinit var providerService: ProviderService
5.     // 测试接口
6.     @GetMapping("/feignProvide")
7.     fun openProvide(): String {
8.         return providerService.provide()
9.     }
10. }
```

5.1.5　Kotlin 集成 Ribbon 服务调用

新建一个 Maven 子工程：chapter05-eureka-consumer-ribbon，它是一个服务消费方，采用 Ribbon 方式消费 5.1.3 节定义的服务。pom 文件如下：

```xml
1.  <?xml version="1.0" encoding="UTF-8"?>
2.  <project xmlns="http://maven.apache.org/POM/4.0.0"
3.           xmlns:xsi="http://www.w3.org/2001/XMLSchema-instance"
4.           xsi:schemaLocation="http://maven.apache.org/POM/4.0.0
    http://maven.apache.org/xsd/maven-4.0.0.xsd">
5.      <!-- 父 pom -->
6.      <parent>
7.          <artifactId>kotlinspringboot</artifactId>
8.          <groupId>io.kang.kotlinspringboot</groupId>
9.          <version>0.0.1-SNAPSHOT</version>
10.     </parent>
11.     <modelVersion>4.0.0</modelVersion>
12.     <!-- 子工程名 -->
13.     <artifactId>chapter05-eureka-consumer-ribbon</artifactId>
14.     <dependencies>
15.         <!-- Eureka Client 依赖包 -->
16.         <dependency>
17.             <groupId>org.springframework.cloud</groupId>
18.             <artifactId>spring-cloud-starter-netflix-eureka-client</artifactId>
19.             <version>2.2.1.RELEASE</version>
20.         </dependency>
21.         <!-- Spring Boot Web 依赖包 -->
22.         <dependency>
23.             <groupId>org.springframework.boot</groupId>
24.             <artifactId>spring-boot-starter-web</artifactId>
25.             <version>2.2.1.RELEASE</version>
26.         </dependency>
27.         <!-- Spring Cloud Ribbon 依赖包 -->
28.         <dependency>
29.             <groupId>org.springframework.cloud</groupId>
30.             <artifactId>spring-cloud-starter-netflix-ribbon</artifactId>
31.             <version>2.2.1.RELEASE</version>
```

```xml
32.        </dependency>
33.        <dependency>
34.            <groupId>com.fasterxml.jackson.module</groupId>
35.            <artifactId>jackson-module-kotlin</artifactId>
36.        </dependency>
37.        <dependency>
38.            <groupId>org.jetbrains.kotlin</groupId>
39.            <artifactId>kotlin-reflect</artifactId>
40.        </dependency>
41.        <dependency>
42.            <groupId>org.jetbrains.kotlin</groupId>
43.            <artifactId>kotlin-stdlib-jdk8</artifactId>
44.        </dependency>
45.        <dependency>
46.            <groupId>org.jetbrains.kotlinx</groupId>
47.            <artifactId>kotlinx-coroutines-core</artifactId>
48.            <version>1.3.2</version>
49.        </dependency>
50.    </dependencies>
51.    <build>
52.        <sourceDirectory>${project.basedir}/src/main/kotlin</sourceDirectory>
53.        <testSourceDirectory>${project.basedir}/src/test/kotlin</testSourceDirectory>
54.        <plugins>
55.            <plugin>
56.                <groupId>org.springframework.boot</groupId>
57.                <artifactId>spring-boot-maven-plugin</artifactId>
58.            </plugin>
59.            <plugin>
60.                <groupId>org.jetbrains.kotlin</groupId>
61.                <artifactId>kotlin-maven-plugin</artifactId>
62.                <configuration>
63.                    <args>
64.                        <arg>-Xjsr305=strict</arg>
65.                    </args>
66.                    <compilerPlugins>
67.                        <plugin>spring</plugin>
68.                        <plugin>jpa</plugin>
```

```xml
69.            </compilerPlugins>
70.          </configuration>
71.          <dependencies>
72.            <dependency>
73.              <groupId>org.jetbrains.kotlin</groupId>
74.              <artifactId>kotlin-maven-allopen</artifactId>
75.              <version>${kotlin.version}</version>
76.            </dependency>
77.            <dependency>
78.              <groupId>org.jetbrains.kotlin</groupId>
79.              <artifactId>kotlin-maven-noarg</artifactId>
80.              <version>${kotlin.version}</version>
81.            </dependency>
82.          </dependencies>
83.        </plugin>
84.      </plugins>
85.    </build>
86. </project>
```

application.yml 的内容如下：

```yaml
1. server:
2.   port: 8002   #应用端口号
3. spring:
4.   application:
5.     name: consumer-ribbon   #应用名称
6. eureka:
7.   client:
8.     service-url:   #Eureka Server 访问地址
9.       defaultZone: http://localhost:8761/eureka/
```

RibbonApplication.kt 定义了启动类，定义一个 restTemplate 和轮询策略，当它启动后会注册到 Eureka Server：

```kotlin
1. // 开启 Eureka Client 注解
2. @SpringBootApplication
3. @EnableEurekaClient
4. class RibbonApplication {
```

```
5.      // 创建 restTemplate，进行负载均衡
6.      @Bean
7.      @LoadBalanced
8.      fun restTemplate(): RestTemplate {
9.          return RestTemplate()
10.     }
11.     // 负载均衡采用随机分配规则
12.     @Bean
13.     fun ribbonRule(): IRule {
14.         return RandomRule()
15.     }
16. }
17. // 启动类
18. fun main(args: Array<String>) {
19.     runApplication<RibbonApplication>(*args)
20. }
```

RibbonService.kt 定义了一个方法，采用 Ribbon 方式访问 chapter05-eureka-provider 定义的接口：

```
1.  @Component
2.  class RibbonService {
3.
4.      @Autowired
5.      lateinit var restTemplate: RestTemplate
6.      // 采用 Ribbon 方式调用 provider-server 的 provide 接口
7.      fun ribbonProvide(): String? {
8.          return restTemplate.getForObject("http://PROVIDER-SERVER/provide", String::class.java)
9.      }
10. }
```

RibbonController.kt 定义了一个接口，用于测试访问服务提供者的接口：

```
1.  @RestController
2.  class RibbonController {
3.      @Autowired
4.      lateinit var ribbonService: RibbonService
```

```
5.        // 测试接口
6.        @GetMapping("/ribbonProvide")
7.        fun ribbonProvide(): String? {
8.            return ribbonService.ribbonProvide()
9.        }
10. }
```

依次启动 chapter05-eureka、chapter05-eureka-provider、chapter05-eureka-consumer、chapter05-eureka-consumer-ribbon，可以看到如图 5.2 所示的服务列表。

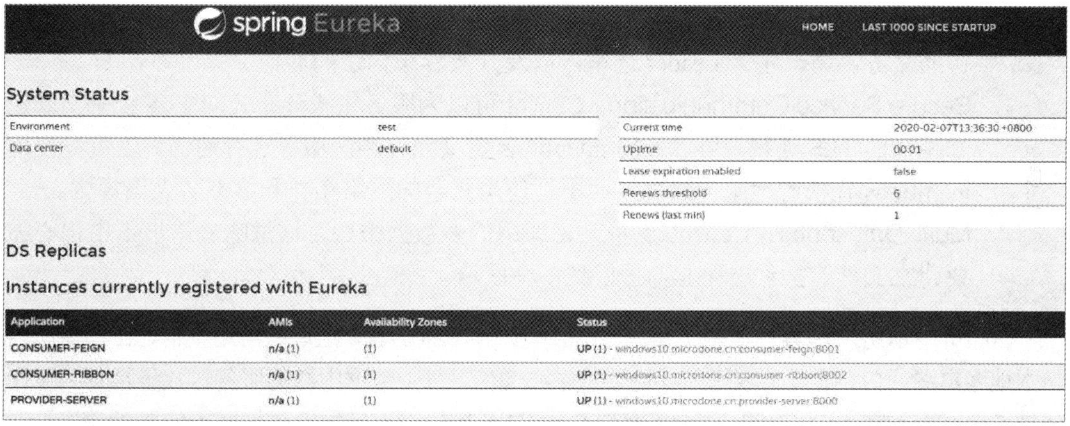

图5.2　Eureka Server注册的服务列表

一个服务提供方和两个服务消费方都注册到了 Eureka Server。调用 consumer-feign 的接口"/feignProvide"和 consumer-ribbon 的接口"/ribbonProvide"，都能访问到 provider-server 的接口"/provide"。

5.2　Consul

Consul 是 HashiCorp 公司推出的开源产品，用于实现分布式系统的服务发现、服务隔离、服务配置。Consul 内置了服务注册与发现框架、分布一致性协议实现、健康检查、Key-Value 存储、多数据中心方案，不再需要依赖其他工具。Consul 本身使用 Go 语言开发，具有跨平台、运行高效等特点。本节介绍使用 Consul 作为 Kotlin 开发的微服务的注册中心的相关知识。

5.2.1 Consul 介绍

Consul 的主要特点有如下几点。

- **Service Discovery**：服务注册与发现，Consul 的客户端可以作为一个服务注册到 Consul，也可以通过 Consul 来查找特定的服务提供者，并且根据提供的信息进行调用。
- **Health Checking**：Consul 客户端会定期发送一些健康检查数据和服务端进行通信，判断客户端的状态、内存使用情况是否正常，用来监控整个集群的状态，防止将请求转给有故障的服务。
- **KV Store**：Consul 还提供了一个容易使用的键值存储。这可以用来保持动态配置、协助服务协调、建立 Leader 选举，以及开发者想构造的其他一些事务。
- **Secure Service Communication**：Consul 可以为服务生成分布式的 TLS 证书，以建立相互的 TLS 连接。可以使用 intentions 定义允许哪些服务进行通信，还可以使用 intentions 轻松管理服务隔离，而不是使用复杂的网络拓扑和静态防火墙规则。
- **Multi Datacenter**：Consul 支持开箱即用的多数据中心，这意味着用户不用担心需要建立额外的抽象层让业务扩展到多个区域。

Consul 有服务端和客户端两个角色，服务端保存配置信息，组成高可用集群，在局域网内与本地客户端通信，通过广域网与其他数据中心通信。每个数据中心的服务器数量推荐为 3 个或是 5 个。客户端是无状态的，负责将 HTTP 和 DNS 接口请求转发给局域网内的服务端集群。

Consul 的调用过程如图 5.3 所示。

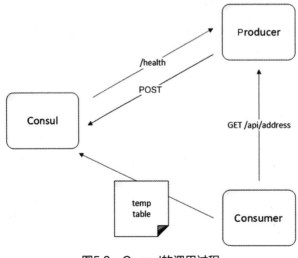

图5.3　Consul的调用过程

当 Producer 启动的时候，会向 Consul 发送一个 POST 请求，告诉 Consul 自己的 IP 地址、端口；Consul 接收到 Producer 的注册后，每隔 10 秒（默认）向 Producer 发送一个健康检查的请求，检验 Producer 是否健康；当 Consumer 发送 GET 方式的请求/api/address 到 Producer 时，会先从 Consul 中拿到一个存储服务 IP 地址、端口的临时表，从表中拿到 Producer 的 IP 地址和端口后再发送 GET 方式的请求/api/address；该临时表每隔 10 秒更新一次，只包含通过健康检查的 Producer。

5.2.2　Kotlin 集成 Consul 服务注册

在 Consul 官网下载 Consul（Consul_1.6.3_windows_amd64），解压后运行 Consul agent – dev 以启动 Consul。

新建一个 Maven 子工程 chapter05-Consul，它是一个服务提供者，也是一个 Consul 客户端。pom 文件如下：

```xml
1.  <?xml version="1.0" encoding="UTF-8"?>
2.  <project xmlns="http://maven.apache.org/POM/4.0.0"
3.           xmlns:xsi="http://www.w3.org/2001/XMLSchema-instance"
4.           xsi:schemaLocation="http://maven.apache.org/POM/4.0.0 http://maven.apache.org/xsd/maven-4.0.0.xsd">
5.      <parent>
6.          <artifactId>kotlinspringboot</artifactId>
7.          <groupId>io.kang.kotlinspringboot</groupId>
8.          <version>0.0.1-SNAPSHOT</version>
9.      </parent>
10.     <modelVersion>4.0.0</modelVersion>
11.     <!-- 子工程名-->
12.     <artifactId>chapter05-consul</artifactId>
13.
14.     <dependencies>
15.         <!-- Spring Cloud Consul 依赖包-->
16.         <dependency>
17.             <groupId>org.springframework.cloud</groupId>
18.             <artifactId>spring-cloud-starter-consul-discovery</artifactId>
19.             <version>2.2.1.RELEASE</version>
20.         </dependency>
21.         <-- Spring Boot Web 依赖包-->
```

```xml
22.     <dependency>
23.         <groupId>org.springframework.boot</groupId>
24.         <artifactId>spring-boot-starter-web</artifactId>
25.         <version>2.2.1.RELEASE</version>
26.     </dependency>
27.     <!-- Spring Boot Actuator 依赖包-->
28.     <dependency>
29.         <groupId>org.springframework.boot</groupId>
30.         <artifactId>spring-boot-starter-actuator</artifactId>
31.         <version>2.2.1.RELEASE</version>
32.     </dependency>
33.     <dependency>
34.         <groupId>com.fasterxml.jackson.module</groupId>
35.         <artifactId>jackson-module-kotlin</artifactId>
36.     </dependency>
37.     <dependency>
38.         <groupId>org.jetbrains.kotlin</groupId>
39.         <artifactId>kotlin-reflect</artifactId>
40.     </dependency>
41.     <dependency>
42.         <groupId>org.jetbrains.kotlin</groupId>
43.         <artifactId>kotlin-stdlib-jdk8</artifactId>
44.     </dependency>
45.     <dependency>
46.         <groupId>org.jetbrains.kotlinx</groupId>
47.         <artifactId>kotlinx-coroutines-core</artifactId>
48.         <version>1.3.2</version>
49.     </dependency>
50. </dependencies>
51. <build>
52.     <sourceDirectory>${project.basedir}/src/main/kotlin</sourceDirectory>
53.     <testSourceDirectory>${project.basedir}/src/test/kotlin</testSourceDirectory>
54.     <plugins>
55.         <plugin>
56.             <groupId>org.springframework.boot</groupId>
57.             <artifactId>spring-boot-maven-plugin</artifactId>
58.         </plugin>
```

```
59.            <plugin>
60.                <groupId>org.jetbrains.kotlin</groupId>
61.                <artifactId>kotlin-maven-plugin</artifactId>
62.                <configuration>
63.                    <args>
64.                        <arg>-Xjsr305=strict</arg>
65.                    </args>
66.                    <compilerPlugins>
67.                        <plugin>spring</plugin>
68.                        <plugin>jpa</plugin>
69.                    </compilerPlugins>
70.                </configuration>
71.                <dependencies>
72.                    <dependency>
73.                        <groupId>org.jetbrains.kotlin</groupId>
74.                        <artifactId>kotlin-maven-allopen</artifactId>
75.                        <version>${kotlin.version}</version>
76.                    </dependency>
77.                    <dependency>
78.                        <groupId>org.jetbrains.kotlin</groupId>
79.                        <artifactId>kotlin-maven-noarg</artifactId>
80.                        <version>${kotlin.version}</version>
81.                    </dependency>
82.                </dependencies>
83.            </plugin>
84.        </plugins>
85.    </build>
86. </project>
```

application.yml 文件的内容如下:

```
1. server:
2.   port: 8080
3. spring:
4.   cloud:
5.     consul:
6.       host: localhost   # consul 主机
7.       port: 8500        # consul 端口
```

```yaml
8.     discovery:    # 注册到 Consul 的服务名
9.       service-name: ${spring.application.name}
10.  application:
11.    name: consul-producer
```

ConsulApplication.kt 定义了一个启动类，启动后会注册到 Consul：

```kotlin
1. // 开启注解，服务注册到 Consul
2. @SpringBootApplication
3. @EnableDiscoveryClient
4. class ConsulApplication
5. // 启动函数
6. fun main(args: Array<String>) {
7.     runApplication<ConsulApplication>(*args)
8. }
```

ConsulController.kt 定义了一个服务接口，用于测试：

```kotlin
1. @RestController
2. class ConsulController {
3.     // 测试接口
4.     @GetMapping("hello/consul")
5.     fun helloConsul(): String {
6.         return "Hello Consul"
7.     }
8. }
```

Consul 的监控界面如图 5.4 所示。

图5.4　Consul的监控界面

5.2.3　Kotlin 集成 OpenFeign 和 Ribbon 服务调用

新建一个 Maven 工程：chapter05-consul-consumer，它是一个消费者，采用 Feign、Ribbon 方式调用 5.2.2 节定义的服务提供者。pom 文件的内容如下：

```xml
1.  <?xml version="1.0" encoding="UTF-8"?>
2.  <project xmlns="http://maven.apache.org/POM/4.0.0"
3.           xmlns:xsi="http://www.w3.org/2001/XMLSchema-instance"
4.           xsi:schemaLocation="http://maven.apache.org/POM/4.0.0
    http://maven.apache.org/xsd/maven-4.0.0.xsd">
5.      <parent>
6.          <artifactId>kotlinspringboot</artifactId>
7.          <groupId>io.kang.kotlinspringboot</groupId>
8.          <version>0.0.1-SNAPSHOT</version>
9.      </parent>
10.     <modelVersion>4.0.0</modelVersion>
11.     <!-- 子工程名 -->
12.     <artifactId>chapter05-consul-consumer</artifactId>
13.     <dependencies>
14.         <!-- Spring Cloud Consul 依赖包-->
15.         <dependency>
16.             <groupId>org.springframework.cloud</groupId>
17.             <artifactId>spring-cloud-starter-consul-discovery</artifactId>
18.             <version>2.2.1.RELEASE</version>
19.         </dependency>
20.         <!-- Spring Cloud OpenFeign 依赖包-->
21.         <dependency>
22.             <groupId>org.springframework.cloud</groupId>
23.             <artifactId>spring-cloud-starter-openfeign</artifactId>
24.             <version>2.2.1.RELEASE</version>
25.         </dependency>
26.         <!-- Spring Boot Web 依赖包-->
27.         <dependency>
28.             <groupId>org.springframework.boot</groupId>
29.             <artifactId>spring-boot-starter-web</artifactId>
30.             <version>2.2.1.RELEASE</version>
31.         </dependency>
```

```xml
32.        <!-- Spring Boot Actuator 依赖包-->
33.        <dependency>
34.            <groupId>org.springframework.boot</groupId>
35.            <artifactId>spring-boot-starter-actuator</artifactId>
36.            <version>2.2.1.RELEASE</version>
37.        </dependency>
38.        <dependency>
39.            <groupId>com.fasterxml.jackson.module</groupId>
40.            <artifactId>jackson-module-kotlin</artifactId>
41.        </dependency>
42.        <dependency>
43.            <groupId>org.jetbrains.kotlin</groupId>
44.            <artifactId>kotlin-reflect</artifactId>
45.        </dependency>
46.        <dependency>
47.            <groupId>org.jetbrains.kotlin</groupId>
48.            <artifactId>kotlin-stdlib-jdk8</artifactId>
49.        </dependency>
50.        <dependency>
51.            <groupId>org.jetbrains.kotlinx</groupId>
52.            <artifactId>kotlinx-coroutines-core</artifactId>
53.            <version>1.3.2</version>
54.        </dependency>
55.    </dependencies>
56.    <build>
57.        <sourceDirectory>${project.basedir}/src/main/kotlin</sourceDirectory>
58.        <testSourceDirectory>${project.basedir}/src/test/kotlin</testSourceDirectory>
59.        <plugins>
60.            <plugin>
61.                <groupId>org.springframework.boot</groupId>
62.                <artifactId>spring-boot-maven-plugin</artifactId>
63.            </plugin>
64.            <plugin>
65.                <groupId>org.jetbrains.kotlin</groupId>
66.                <artifactId>kotlin-maven-plugin</artifactId>
67.                <configuration>
68.                    <args>
```

```
69.                    <arg>-Xjsr305=strict</arg>
70.                </args>
71.                <compilerPlugins>
72.                    <plugin>spring</plugin>
73.                    <plugin>jpa</plugin>
74.                </compilerPlugins>
75.            </configuration>
76.            <dependencies>
77.                <dependency>
78.                    <groupId>org.jetbrains.kotlin</groupId>
79.                    <artifactId>kotlin-maven-allopen</artifactId>
80.                    <version>${kotlin.version}</version>
81.                </dependency>
82.                <dependency>
83.                    <groupId>org.jetbrains.kotlin</groupId>
84.                    <artifactId>kotlin-maven-noarg</artifactId>
85.                    <version>${kotlin.version}</version>
86.                </dependency>
87.            </dependencies>
88.        </plugin>
89.    </plugins>
90. </build>
91. </project>
```

application.yml 文件的内容如下所示：

```
1. spring:
2.   application:
3.     name: consul-consumer    #应用名称
4.   cloud:
5.     consul:
6.       host: 127.0.0.1    #Consul 主机
7.       port: 8500    #Consul 端口
8.       discovery:
9.         register: false  # 仅作为消费者,不注册服务
10. server:
11.   port: 8005    #应用端口号
```

ConsulConsumer.kt 定义了一个启动类，启动后运行一个 Consul 消费者：

```kotlin
1.  // 开启 Feign 注解，开启服务注册注解
2.  @SpringBootApplication
3.  @EnableDiscoveryClient
4.  @EnableFeignClients
5.  class ConsulConsumer {
6.      @Bean
7.      @LoadBalanced
8.      fun restTemplate(): RestTemplate {
9.          return RestTemplate()
10.     }
11. }
12. // 启动函数
13. fun main(args: Array<String>) {
14.     runApplication<ConsulConsumer>(*args)
15. }
```

FeignService.kt 定义了一个 Feign 接口，用于调用"hello/consul"这个接口：

```kotlin
1.  @FeignClient(value = "consul-producer")
2.  interface FeignService {
3.      // 测试接口，通过 Feign 方式调用
4.      @GetMapping("hello/consul")
5.      fun helloConsul(): String
6.  }
```

ConsumerController.kt 定义了 Feign 和 Ribbon 的测试接口：

```kotlin
1.  @RestController
2.  class ConsumerController {
3.      @Autowired
4.      lateinit var restTemplate: RestTemplate
5.      @Autowired
6.      lateinit var feignService: FeignService
7.      // 测试接口，通过 Ribbon 方式调用
8.      @GetMapping("ribbon/hello/consul")
9.      fun ribbonHelloConsul(): String? {
```

```
10.         return restTemplate.getForObject("http://consul-producer/hello/consul",
    String::class.java)
11.     }
12.     // 测试接口,通过 Feign 方式调用
13.     @GetMapping("feign/hello/consul")
14.     fun feignHelloConsul(): String {
15.         return feignService.helloConsul()
16.     }
17. }
```

服务启动后,通过"ribbon/hello/consul"采用 Ribbon 方式可以访问"hello/consul"这个服务接口。通过"feign/hello/consul"采用 Feign 方式也可以访问"hello/consul"这个服务接口。

5.3 Zookeeper

Zookeeper 是一个分布式的、开源的程序协调服务,是 Hadoop 项目下的一个子项目。它提供的主要功能包括:配置管理、名字服务、分布式锁、集群管理。本节介绍使用 Zookeeper 作为 Kotlin 开发的微服务的注册中心的相关知识。

5.3.1 Zookeeper 介绍

Zookeeper 使用 Zab 协议来提供一致性。很多开源项目使用 Zookeeper 来维护配置,比如,HBase 的客户端通过 Zookeeper 获得必要的 HBase 集群的配置信息,开源消息队列 Kafka 使用 Zookeeper 维护 Broker 的信息,Alibaba 开源的 SOA 框架 Dubbo 使用 Zookeeper 管理一些配置来实现服务治理。

Zookeeper 的命名服务功能主要是根据指定名字来获取资源或服务的地址、提供者等信息,利用其 znode 的特点和 watcher 机制,将其作为动态注册和获取服务信息的配置中心,统一管理服务名称和其对应的服务器列表信息,能够近乎实时地感知到后端服务器的状态(上线、下线、宕机)。

在分布式环境中,为了提高可靠性,集群中的每台服务器上都部署着同样的服务。集群中的每台服务器需要协调,使用分布式锁,在某个时刻只让一台服务器工作,当这台服务器出问题的时候释放锁。首先需要创建一个父节点,尽量是持久节点,然后每个要获得锁的线程都会在这个节点下创建一个临时顺序节点,由于序号的递增性,可以规定排号最

小的那个获得锁。Zookeeper 的节点监听机制可以保障占有锁的方式有序而且高效。

Zookeeper 可以管理集群和服务发现。在分布式的集群中，经常会由于各种原因，比如硬件故障、软件故障、网络问题，有些节点会进进出出。有新的节点加入进来，也有老的节点退出集群。这个时候，集群中的其他机器需要能感知到这种变化，然后根据这种变化做出对应的决策。当消费者访问某个服务时，就需要采用某种机制发现现在有哪些节点可以提供该服务。

在生产环境中，ZooKeeper 采用集群部署。Zookeeper 集群中包含 Leader、Follower 以及 Observer 三个角色。Leader 负责进行投票的发起和决议，更新系统状态，Leader 由选举产生；Follower 用于接收客户端请求并向客户端返回结果，在选举过程中参与投票；Observer 可以接收客户端连接，接收读写请求，将写请求转发给 Leader，但 Observer 不参加投票过程，只同步 Leader 的状态，Observer 的目的是为了扩展系统，提高读取速度。在 Zookeeper 系统中，只要集群中存在超过一半的节点（这里指的是投票节点即非 Observer 节点）能够正常工作，那么整个集群就能够正常对外服务。

Zookeeper 的集群架构如图 5.5 所示。

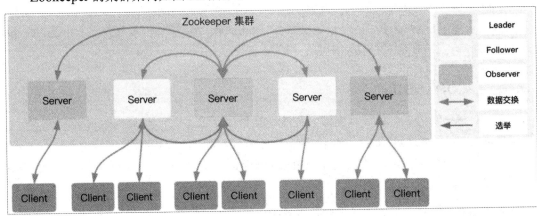

图5.5　Zookeeper的集群架构

5.3.2　Kotlin 集成 Zookeeper 服务注册

新建一个子工程：chapter05-zk，这是一个服务提供者，pom 文件如下：

```
1.  <?xml version="1.0" encoding="UTF-8"?>
2.  <project xmlns="http://maven.apache.org/POM/4.0.0"
3.          xmlns:xsi="http://www.w3.org/2001/XMLSchema-instance"
```

```xml
4.        xsi:schemaLocation="http://maven.apache.org/POM/4.0.0
    http://maven.apache.org/xsd/maven-4.0.0.xsd">
5.    <parent>
6.        <artifactId>kotlinspringboot</artifactId>
7.        <groupId>io.kang.kotlinspringboot</groupId>
8.        <version>0.0.1-SNAPSHOT</version>
9.    </parent>
10.   <modelVersion>4.0.0</modelVersion>
11.   <!-- 子工程名 -->
12.   <artifactId>chapter05-zk</artifactId>
13.   <dependencies>
14.       <!-- Spring Cloud Zookeeper 依赖包 -->
15.       <dependency>
16.           <groupId>org.springframework.cloud</groupId>
17.           <artifactId>spring-cloud-starter-zookeeper-discovery</artifactId>
18.           <version>2.2.0.RELEASE</version>
19.       </dependency>
20.       <!-- Spring Boot Web 依赖包 -->
21.       <dependency>
22.           <groupId>org.springframework.boot</groupId>
23.           <artifactId>spring-boot-starter-web</artifactId>
24.           <version>2.2.1.RELEASE</version>
25.       </dependency>
26.       <dependency>
27.           <groupId>com.fasterxml.jackson.module</groupId>
28.           <artifactId>jackson-module-kotlin</artifactId>
29.       </dependency>
30.       <dependency>
31.           <groupId>org.jetbrains.kotlin</groupId>
32.           <artifactId>kotlin-reflect</artifactId>
33.       </dependency>
34.       <dependency>
35.           <groupId>org.jetbrains.kotlin</groupId>
36.           <artifactId>kotlin-stdlib-jdk8</artifactId>
37.       </dependency>
38.       <dependency>
39.           <groupId>org.jetbrains.kotlinx</groupId>
```

```xml
40.             <artifactId>kotlinx-coroutines-core</artifactId>
41.             <version>1.3.2</version>
42.         </dependency>
43.     </dependencies>
44.     <build>
45.         <sourceDirectory>${project.basedir}/src/main/kotlin</sourceDirectory>
46.         <testSourceDirectory>${project.basedir}/src/test/kotlin</testSourceDirectory>
47.         <plugins>
48.             <plugin>
49.                 <groupId>org.springframework.boot</groupId>
50.                 <artifactId>spring-boot-maven-plugin</artifactId>
51.             </plugin>
52.             <plugin>
53.                 <groupId>org.jetbrains.kotlin</groupId>
54.                 <artifactId>kotlin-maven-plugin</artifactId>
55.                 <configuration>
56.                     <args>
57.                         <arg>-Xjsr305=strict</arg>
58.                     </args>
59.                     <compilerPlugins>
60.                         <plugin>spring</plugin>
61.                         <plugin>jpa</plugin>
62.                     </compilerPlugins>
63.                 </configuration>
64.                 <dependencies>
65.                     <dependency>
66.                         <groupId>org.jetbrains.kotlin</groupId>
67.                         <artifactId>kotlin-maven-allopen</artifactId>
68.                         <version>${kotlin.version}</version>
69.                     </dependency>
70.                     <dependency>
71.                         <groupId>org.jetbrains.kotlin</groupId>
72.                         <artifactId>kotlin-maven-noarg</artifactId>
73.                         <version>${kotlin.version}</version>
74.                     </dependency>
75.                 </dependencies>
76.             </plugin>
```

```
77.        </plugins>
78.    </build>
79. </project>
```

application.yml 文件的内容如下所示：

```
1. server:
2.   port: 8081   #应用端口号
3. spring:
4.   application:
5.     name: zk-produce   #应用名称
6.   cloud:
7.     zookeeper:   #Zookeeper 的主机、端口
8.       connect-string: localhost:2181
```

ZkApplication.kt 定义了一个启动类，服务启动后，会注册到 Zookeeper：

```
1. // 开启服务注册注解
2. @SpringBootApplication
3. @EnableDiscoveryClient
4. class ZkApplication
5. // 启动函数
6. fun main(args: Array<String>) {
7.     runApplication<ZkApplication>(*args)
8. }
```

服务在 Zookeeper 中的注册信息如下所示，包括应用名、id、address、port 等信息：

```
1.  {
2.      "name": "zk-produce",
3.      "id": "133ff0ac-cd6f-4942-886c-b76243df7550",
4.      "address": "windows10.microdone.cn",
5.      "port": 8081,
6.      "sslPort": null,
7.      "payload": {
8.          "@class": "org.springframework.cloud.zookeeper.discovery.ZookeeperInstance",
9.          "id": "application-1",
10.         "name": "zk-produce",
11.         "metadata": {}
```

```
12.        },
13.        "registrationTimeUTC": 1581241617714,
14.        "serviceType": "DYNAMIC",
15.        "uriSpec": {
16.            "parts": [{
17.                "value": "scheme",
18.                "variable": true
19.            }, {
20.                "value": "://",
21.                "variable": false
22.            }, {
23.                "value": "address",
24.                "variable": true
25.            }, {
26.                "value": ":",
27.                "variable": false
28.            }, {
29.                "value": "port",
30.                "variable": true
31.            }]
32.        }
33. }
```

ZkController.kt 定义了一个测试接口：

```
1. @RestController
2. class ZkController {
3.     // 测试接口
4.     @GetMapping("hello/zk")
5.     fun helloZk(): String {
6.         return "hello zookeeper"
7.     }
8. }
```

5.3.3　Kotlin 集成 OpenFeign 和 Ribbon 服务调用

新建一个子工程：chapter05-zk-consumer，这是一个服务消费者，pom 文件如下：

```xml
1.  <?xml version="1.0" encoding="UTF-8"?>
2.  <project xmlns="http://maven.apache.org/POM/4.0.0"
3.           xmlns:xsi="http://www.w3.org/2001/XMLSchema-instance"
4.           xsi:schemaLocation="http://maven.apache.org/POM/4.0.0 http://maven.apache.org/xsd/maven-4.0.0.xsd">
5.      <parent>
6.          <artifactId>kotlinspringboot</artifactId>
7.          <groupId>io.kang.kotlinspringboot</groupId>
8.          <version>0.0.1-SNAPSHOT</version>
9.      </parent>
10.     <modelVersion>4.0.0</modelVersion>
11.     <!-- 子工程名 -->
12.     <artifactId>chapter05-zk-consumer</artifactId>
13.     <dependencies>
14.         <!-- Spring Cloud Zookeeper 依赖包 -->
15.         <dependency>
16.             <groupId>org.springframework.cloud</groupId>
17.             <artifactId>spring-cloud-starter-zookeeper-discovery</artifactId>
18.             <version>2.2.0.RELEASE</version>
19.         </dependency>
20.         <!-- Spring Boot Web 依赖包 -->
21.         <dependency>
22.             <groupId>org.springframework.boot</groupId>
23.             <artifactId>spring-boot-starter-web</artifactId>
24.             <version>2.2.1.RELEASE</version>
25.         </dependency>
26.         <!-- Spring Cloud OpenFeign 依赖包 -->
27.         <dependency>
28.             <groupId>org.springframework.cloud</groupId>
29.             <artifactId>spring-cloud-starter-openfeign</artifactId>
30.             <version>2.2.1.RELEASE</version>
31.         </dependency>
32.         <dependency>
33.             <groupId>com.fasterxml.jackson.module</groupId>
34.             <artifactId>jackson-module-kotlin</artifactId>
35.         </dependency>
```

```xml
36.     <dependency>
37.         <groupId>org.jetbrains.kotlin</groupId>
38.         <artifactId>kotlin-reflect</artifactId>
39.     </dependency>
40.     <dependency>
41.         <groupId>org.jetbrains.kotlin</groupId>
42.         <artifactId>kotlin-stdlib-jdk8</artifactId>
43.     </dependency>
44.     <dependency>
45.         <groupId>org.jetbrains.kotlinx</groupId>
46.         <artifactId>kotlinx-coroutines-core</artifactId>
47.         <version>1.3.2</version>
48.     </dependency>
49. </dependencies>
50. <build>
51.     <sourceDirectory>${project.basedir}/src/main/kotlin</sourceDirectory>
52.     <testSourceDirectory>${project.basedir}/src/test/kotlin</testSourceDirectory>
53.     <plugins>
54.         <plugin>
55.             <groupId>org.springframework.boot</groupId>
56.             <artifactId>spring-boot-maven-plugin</artifactId>
57.         </plugin>
58.         <plugin>
59.             <groupId>org.jetbrains.kotlin</groupId>
60.             <artifactId>kotlin-maven-plugin</artifactId>
61.             <configuration>
62.                 <args>
63.                     <arg>-Xjsr305=strict</arg>
64.                 </args>
65.                 <compilerPlugins>
66.                     <plugin>spring</plugin>
67.                     <plugin>jpa</plugin>
68.                 </compilerPlugins>
69.             </configuration>
70.             <dependencies>
71.                 <dependency>
72.                     <groupId>org.jetbrains.kotlin</groupId>
```

```xml
73.                <artifactId>kotlin-maven-allopen</artifactId>
74.                <version>${kotlin.version}</version>
75.            </dependency>
76.            <dependency>
77.                <groupId>org.jetbrains.kotlin</groupId>
78.                <artifactId>kotlin-maven-noarg</artifactId>
79.                <version>${kotlin.version}</version>
80.            </dependency>
81.         </dependencies>
82.      </plugin>
83.    </plugins>
84. </build>
85. </project>
```

application.yml 文件的内容如下：

```yml
1. server:
2.   port: 8006    #应用端口号
3. spring:
4.   application:
5.     name: zk-consumer    #应用名称
6.   cloud:
7.     zookeeper:    #Zookeeper 连接地址
8.       connect-string: localhost:2181
```

ZkConsumerApplication.kt 定义了启动类，服务启动后，会注册到 Zookeeper：

```kotlin
1. //开启 Feign 注解，开启服务注册注解
2. @EnableDiscoveryClient
3. @SpringBootApplication
4. @EnableFeignClients
5. class ZkConsumerApplication {
6.     @Bean
7.     @LoadBalanced
8.     fun restTemplate():RestTemplate {
9.         return RestTemplate()
10.    }
11. }
```

```
12.  // 启动函数
13.  fun main(args: Array<String>) {
14.      runApplication<ZkConsumerApplication>(*args)
15.  }
```

FeignService.kt 定义了一个测试接口,使用 Feign 方式调用 zk-produce 的 hello/zk 接口:

```
1.  @FeignClient(value = "zk-produce", path = "/")
2.  @Component
3.  interface FeignService {
4.      // 测试接口,使用 feign 方式
5.      @GetMapping("hello/zk")
6.      fun helloZk(): String
7.  }
```

ZkConsumerController.kt 定义了两个测试接口,分别用 Ribbon、Feign 方式调用 zk-produce 的 hello/zk 接口:

```
1.  @RestController
2.  class ZkConsumerController {
3.      @Autowired
4.      lateinit var feignService: FeignService
5.
6.      @Autowired
7.      lateinit var restTemplate: RestTemplate
8.      // 测试接口,使用 Feign 方式
9.      @GetMapping("feign/hello/zk")
10.     fun feignHelloZk(): String {
11.         return feignService.helloZk()
12.     }
13.     // 测试接口,使用 Ribbon 方式
14.     @GetMapping("ribbon/hello/zk")
15.     fun ribbonHelloZk(): String? {
16.         return restTemplate.getForObject("http://zk-produce/hello/zk", String::class.java)
17.     }
18. }
```

5.4 Nacos

Nacos 是阿里巴巴开源的服务发现、配置和管理工具。Nacos 提供了一组简单易用的特性集，有助于快速实现动态服务发现、服务配置、服务元数据及流量管理。本节介绍使用 Nacos 作为 Kotlin 开发的微服务注册中心的相关知识。

5.4.1 Nacos 介绍

Nacos 的特性有如下几点。

服务发现和服务健康监测：Nacos 支持基于 DNS 和基于 RPC 的服务发现。服务提供者使用原生 SDK、OpenAPI 或一个独立的 Agent 注册 Service 后，服务消费者可以使用 DNS 或 HTTP&API 查找和发现服务。Nacos 提供对服务的实时的健康检查，阻止向不健康的主机或服务实例发送请求。Nacos 支持传输层（PING 或 TCP）和应用层（如 HTTP、MySQL、用户自定义）的健康检查。对于复杂的云环境和网络拓扑环境（如 VPC、边缘网络等）中的服务的健康检查，Nacos 提供了 Agent 上报模式和服务端主动检测两种健康检查模式。Nacos 还提供了统一的健康检查仪表盘，可根据健康状态管理服务的可用性及流量。

动态配置服务：动态配置服务有助于以中心化、外部化和动态化的方式管理所有环境的应用配置和服务配置。动态配置消除了配置变更时重新部署应用和服务的需要，让配置管理变得更加高效和敏捷。配置中心化管理让实现无状态服务变得更简单，让服务按需弹性伸缩变得更容易。Nacos 提供了一个简单易用的 UI 管理所有服务和应用的配置。Nacos 还提供包括配置版本跟踪、金丝雀发布、一键回滚配置以及客户端配置更新状态跟踪在内的一系列开箱即用的配置管理特性，使人们可以更安全地在生产环境中管理配置变更和降低配置变更带来的风险。

动态 DNS 服务：动态 DNS 服务支持权重路由，可更容易地实现中间层负载均衡、更灵活的路由策略、流量控制以及数据中心内网的简单 DNS 解析服务。动态 DNS 服务可以实现以 DNS 协议为基础的服务发现，消除耦合到厂商私有服务发现 API 上的风险。Nacos 还提供了一些简单的 DNS API 管理服务的关联域名和可用的 IP:PORT 列表。

服务及其元数据管理：Nacos 从微服务平台建设的角度管理数据中心的所有服务及元数据，包括管理服务的描述、服务的生命周期、服务的静态依赖分析、服务的健康状态、服务的流量管理、路由及安全策略、服务的 SLA 以及指标统计数据。

5.4.2　Kotlin 集成 Nacos 服务注册

新建一个 Maven 子工程：chapter05-nacos，这是一个服务提供者，服务启动后可以注册到 Nacos。pom 文件如下：

```
1.  <?xml version="1.0" encoding="UTF-8"?>
2.  <project xmlns="http://maven.apache.org/POM/4.0.0"
3.      xmlns:xsi="http://www.w3.org/2001/XMLSchema-instance"
4.      xsi:schemaLocation="http://maven.apache.org/POM/4.0.0
    http://maven.apache.org/xsd/maven-4.0.0.xsd">
5.      <parent>
6.          <artifactId>kotlinspringboot</artifactId>
7.          <groupId>io.kang.kotlinspringboot</groupId>
8.          <version>0.0.1-SNAPSHOT</version>
9.      </parent>
10.     <modelVersion>4.0.0</modelVersion>
11.     <!-- 子工程名 -->
12.     <artifactId>chapter05-nacos</artifactId>
13.     <dependencies>
14.         <!-- Spring Cloud Nacos 依赖包 -->
15.         <dependency>
16.             <groupId>com.alibaba.cloud</groupId>
17.             <artifactId>spring-cloud-starter-alibaba-nacos-discovery</artifactId>
18.             <version>2.1.1.RELEASE</version>
19.         </dependency>
20.         <!-- Spring Boot Web 依赖包 -->
21.         <dependency>
22.             <groupId>org.springframework.boot</groupId>
23.             <artifactId>spring-boot-starter-web</artifactId>
24.             <version>2.2.1.RELEASE</version>
25.         </dependency>
26.         <dependency>
27.             <groupId>com.fasterxml.jackson.module</groupId>
28.             <artifactId>jackson-module-kotlin</artifactId>
29.         </dependency>
30.         <dependency>
31.             <groupId>org.jetbrains.kotlin</groupId>
```

```xml
32.            <artifactId>kotlin-reflect</artifactId>
33.        </dependency>
34.        <dependency>
35.            <groupId>org.jetbrains.kotlin</groupId>
36.            <artifactId>kotlin-stdlib-jdk8</artifactId>
37.        </dependency>
38.        <dependency>
39.            <groupId>org.jetbrains.kotlinx</groupId>
40.            <artifactId>kotlinx-coroutines-core</artifactId>
41.            <version>1.3.2</version>
42.        </dependency>
43.    </dependencies>
44.    <build>
45.        <sourceDirectory>${project.basedir}/src/main/kotlin</sourceDirectory>
46.        <testSourceDirectory>${project.basedir}/src/test/kotlin</testSourceDirectory>
47.        <plugins>
48.            <plugin>
49.                <groupId>org.springframework.boot</groupId>
50.                <artifactId>spring-boot-maven-plugin</artifactId>
51.            </plugin>
52.            <plugin>
53.                <groupId>org.jetbrains.kotlin</groupId>
54.                <artifactId>kotlin-maven-plugin</artifactId>
55.                <configuration>
56.                    <args>
57.                        <arg>-Xjsr305=strict</arg>
58.                    </args>
59.                    <compilerPlugins>
60.                        <plugin>spring</plugin>
61.                        <plugin>jpa</plugin>
62.                    </compilerPlugins>
63.                </configuration>
64.                <dependencies>
65.                    <dependency>
66.                        <groupId>org.jetbrains.kotlin</groupId>
67.                        <artifactId>kotlin-maven-allopen</artifactId>
68.                        <version>${kotlin.version}</version>
```

```xml
69.            </dependency>
70.            <dependency>
71.                <groupId>org.jetbrains.kotlin</groupId>
72.                <artifactId>kotlin-maven-noarg</artifactId>
73.                <version>${kotlin.version}</version>
74.            </dependency>
75.          </dependencies>
76.        </plugin>
77.      </plugins>
78.    </build>
79. </project>
```

application.yml 文件的内容如下：

```yaml
1. server:
2.   port: 8100   #服务端口号
3. spring:
4.   application:
5.     name: nacos-producer   #服务名称
6.   cloud:
7.     nacos:
8.       discovery:   #Nacos 服务中心地址
9.         server-addr: 127.0.0.1:8848
```

NacosApplication.kt 是一个启动类：

```kotlin
1. // 开启服务注册注解
2. @SpringBootApplication
3. @EnableDiscoveryClient
4. class NacosApplication
5. // 启动函数
6. fun main(args: Array<String>) {
7.     runApplication<NacosApplication>(*args)
8. }
```

NacosController.kt 定义了一个服务接口：

```kotlin
1. @RestController
2. class NacosController {
```

```
3.      // 测试接口
4.      @GetMapping("hello/nacos")
5.      fun helloNacos(): String {
6.          return "Hello Nacos"
7.      }
8.  }
```

5.4.3　Kotlin 集成 OpenFeign 和 Ribbon 服务调用

新建一个 Maven 子工程：chapter05-nacos-consumer，这是一个服务消费方，采用 Feign 和 Ribbon 方式访问 chapter05-nacos 定义的服务接口。pom 文件如下：

```
1.  <?xml version="1.0" encoding="UTF-8"?>
2.  <project xmlns="http://maven.apache.org/POM/4.0.0"
3.      xmlns:xsi="http://www.w3.org/2001/XMLSchema-instance"
4.      xsi:schemaLocation="http://maven.apache.org/POM/4.0.0
    http://maven.apache.org/xsd/maven-4.0.0.xsd">
5.      <parent>
6.          <artifactId>kotlinspringboot</artifactId>
7.          <groupId>io.kang.kotlinspringboot</groupId>
8.          <version>0.0.1-SNAPSHOT</version>
9.      </parent>
10.     <modelVersion>4.0.0</modelVersion>
11.     <!-- 子工程名 -->
12.     <artifactId>chapter05-nacos-consumer</artifactId>
13.     <dependencies>
14.         <!-- Spring Cloud Nacos 依赖包 -->
15.         <dependency>
16.             <groupId>com.alibaba.cloud</groupId>
17.             <artifactId>spring-cloud-starter-alibaba-nacos-discovery</artifactId>
18.             <version>2.1.1.RELEASE</version>
19.         </dependency>
20.         <!-- Spring Boot Web 依赖包 -->
21.         <dependency>
22.             <groupId>org.springframework.boot</groupId>
23.             <artifactId>spring-boot-starter-web</artifactId>
24.             <version>2.2.1.RELEASE</version>
```

```xml
25.        </dependency>
26.        <!-- Spring Cloud OpenFeign 依赖包 -->
27.        <dependency>
28.            <groupId>org.springframework.cloud</groupId>
29.            <artifactId>spring-cloud-starter-openfeign</artifactId>
30.            <version>2.2.1.RELEASE</version>
31.        </dependency>
32.        <dependency>
33.            <groupId>com.fasterxml.jackson.module</groupId>
34.            <artifactId>jackson-module-kotlin</artifactId>
35.        </dependency>
36.        <dependency>
37.            <groupId>org.jetbrains.kotlin</groupId>
38.            <artifactId>kotlin-reflect</artifactId>
39.        </dependency>
40.        <dependency>
41.            <groupId>org.jetbrains.kotlin</groupId>
42.            <artifactId>kotlin-stdlib-jdk8</artifactId>
43.        </dependency>
44.        <dependency>
45.            <groupId>org.jetbrains.kotlinx</groupId>
46.            <artifactId>kotlinx-coroutines-core</artifactId>
47.            <version>1.3.2</version>
48.        </dependency>
49.    </dependencies>
50.    <build>
51.        <sourceDirectory>${project.basedir}/src/main/kotlin</sourceDirectory>
52.        <testSourceDirectory>${project.basedir}/src/test/kotlin</testSourceDirectory>
53.        <plugins>
54.            <plugin>
55.                <groupId>org.springframework.boot</groupId>
56.                <artifactId>spring-boot-maven-plugin</artifactId>
57.            </plugin>
58.            <plugin>
59.                <groupId>org.jetbrains.kotlin</groupId>
60.                <artifactId>kotlin-maven-plugin</artifactId>
61.                <configuration>
```

```xml
62.            <args>
63.                <arg>-Xjsr305=strict</arg>
64.            </args>
65.            <compilerPlugins>
66.                <plugin>spring</plugin>
67.                <plugin>jpa</plugin>
68.            </compilerPlugins>
69.        </configuration>
70.        <dependencies>
71.            <dependency>
72.                <groupId>org.jetbrains.kotlin</groupId>
73.                <artifactId>kotlin-maven-allopen</artifactId>
74.                <version>${kotlin.version}</version>
75.            </dependency>
76.            <dependency>
77.                <groupId>org.jetbrains.kotlin</groupId>
78.                <artifactId>kotlin-maven-noarg</artifactId>
79.                <version>${kotlin.version}</version>
80.            </dependency>
81.        </dependencies>
82.    </plugin>
83.   </plugins>
84.  </build>
85. </project>
```

application.yml 文件的内容如下：

```
1. server:
2.   port: 8006   #应用端口号
3. spring:
4.   application:
5.     name: nacos-consumer   #服务名
6.   cloud:
7.     nacos:
8.       discovery:   #Nacos 注册中心地址
9.         server-addr: 127.0.0.1:8848
```

NacosConsumerApplication.kt 定义了一个启动类：

```kotlin
1.  // 开启服务注册注解，开启Feign注解
2.  @SpringBootApplication
3.  @EnableFeignClients
4.  @EnableDiscoveryClient
5.  class NacosConsumerApplication {
6.      @Bean
7.      @LoadBalanced
8.      fun restTemplate(): RestTemplate {
9.          return RestTemplate()
10.     }
11. }
12. // 启动函数
13. fun main(args: Array<String>) {
14.     runApplication<NacosConsumerApplication>(*args)
15. }
```

FeignService.kt 定义了一个 Feign 接口：

```kotlin
1.  @FeignClient(value = "nacos-producer")
2.  @Component
3.  interface FeignService {
4.      // 测试接口，使用Feign方式调用
5.      @GetMapping("hello/nacos")
6.      fun helloNacos(): String
7.  }
```

NacosConsumerController.kt 定义了两个测试接口，分别用 Ribbon 和 Feign 方式调用服务接口：

```kotlin
1.  @RestController
2.  class NacosConsumerController {
3.      @Autowired
4.      lateinit var restTemplate: RestTemplate
5.      @Autowired
6.      lateinit var feignService: FeignService
7.      // 测试接口，使用Ribbon方式调用
8.      @GetMapping("ribbon/hello/nacos")
```

```kotlin
9.      fun ribbonHelloNacos(): String? {
10.         return restTemplate.getForObject("http://nacos-producer/hello/nacos",
String::class.java)
11.     }
12.     // 测试接口，使用 Feign 方式调用
13.     @GetMapping("feign/hello/nacos")
14.     fun feignHelloNacos(): String {
15.         return feignService.helloNacos()
16.     }
17. }
```

使用 Ribbon 方式通过"ribbon/hello/nacos"或使用 Feign 方式通过"feign/hello/nacos"都可以访问到"hello/nacos"接口。

Nacos 监控页面如图 5.6 所示。

图5.6　Nacos监控界面

5.5　小结

本章介绍了 Kotlin 应用于微服务注册中心的相关内容，通过示例介绍了服务注册和调用方法。服务注册中心对于微服务系统很重要，可影响服务的注册、发现、治理。使用服务注册中心，可将单体应用拆分为微服务，微服务组件通过服务注册中心通信。

第 6 章
Kotlin 应用于微服务配置中心

配置中心可以存储微服务系统的配置，当配置更新时，可以下发到微服务系统。本章将介绍四种配置中心：Spring Cloud Config、Apollo、Nacos 和 Consul。本章将使用示例展示 Kotlin 集成微服务配置中心的具体方法。

6.1 Spring Cloud Config

Spring Cloud Config 是 Spring Cloud 家族的中心组件，负责配置文件的统一管理及实时更新。本节主要介绍 Spring Cloud Config 作为 Kotlin 开发的微服务的配置中心的相关知识。

6.1.1 Spring Cloud Config 介绍

Spring Cloud Config 支持将配置文件放在配置服务的内存中（即本地），也支持放在远程 Git 仓库中。在 Spring Cloud Config 组件中有两种角色，一种是 config server，另一种是 config client。可以通过 RESTful 接口访问配置文件、请求地址和存放资源文件映射，如下所示：

```
/{application}/{profile}[/{label}]
/{application}-{profile}.yml
/{label}/{application}-{profile}.yml
```

/{application}-{profile}.properties
/{label}/{application}-{profile}.properties

application 是应用的名称，profile 区分开发环境、测试环境、生产环境配置文件，label 是配置的分支标签，用于版本管理。Spring Boot 支持 yml 和 properties 两种格式的配置文件。

服务实例都将从配置中心读取文件，可以将配置中心做成一个微服务，将其集群化，从而达到高可用，如图 6.1 所示。

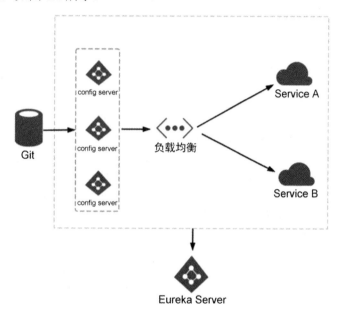

图6.1 服务配置中心示意图

Spring Cloud Config 借助 Spring Cloud Bus 实现配置的热加载。Spring Cloud Bus 会对外提供一个 HTTP 接口，即/bus-refresh。将这个接口配置到远程的 Git 的 webhook 上，当 Git 上的文件内容发生变动时，Git 会自动调用/bus-refresh 接口。Bus 就会通知 config server，config server 会发布更新消息到消息总线的消息队列中，其他服务订阅到该消息就会进行刷新，从而实现整个微服务进行自动刷新。

6.1.2　Kotlin 集成 Spring Cloud Config

新建一个 Maven 子工程：chapter06-springcloud-config，这是一个 config server。pom 文件如下：

```xml
1.  <?xml version="1.0" encoding="UTF-8"?>
2.  <project xmlns="http://maven.apache.org/POM/4.0.0"
3.      xmlns:xsi="http://www.w3.org/2001/XMLSchema-instance"
4.      xsi:schemaLocation="http://maven.apache.org/POM/4.0.0 http://maven.apache.org/xsd/maven-4.0.0.xsd">
5.      <parent>
6.          <artifactId>kotlinspringboot</artifactId>
7.          <groupId>io.kang.kotlinspringboot</groupId>
8.          <version>0.0.1-SNAPSHOT</version>
9.      </parent>
10.     <modelVersion>4.0.0</modelVersion>
11.     <!-- 子工程名 -->
12.     <artifactId>chapter06-springcloud-config</artifactId>
13.     <dependencies>
14.         <!-- Spring Cloud Config Server 依赖包 -->
15.         <dependency>
16.             <groupId>org.springframework.cloud</groupId>
17.             <artifactId>spring-cloud-config-server</artifactId>
18.             <version>2.2.1.RELEASE</version>
19.         </dependency>
20.         <dependency>
21.             <groupId>com.fasterxml.jackson.module</groupId>
22.             <artifactId>jackson-module-kotlin</artifactId>
23.         </dependency>
24.         <dependency>
25.             <groupId>org.jetbrains.kotlin</groupId>
26.             <artifactId>kotlin-reflect</artifactId>
27.         </dependency>
28.         <dependency>
29.             <groupId>org.jetbrains.kotlin</groupId>
30.             <artifactId>kotlin-stdlib-jdk8</artifactId>
31.         </dependency>
32.         <dependency>
33.             <groupId>org.jetbrains.kotlinx</groupId>
34.             <artifactId>kotlinx-coroutines-core</artifactId>
35.             <version>1.3.2</version>
36.         </dependency>
```

```xml
37.        </dependencies>
38.        <build>
39.            <sourceDirectory>${project.basedir}/src/main/kotlin</sourceDirectory>
40.            <testSourceDirectory>${project.basedir}/src/test/kotlin</testSourceDirectory>
41.            <plugins>
42.                <plugin>
43.                    <groupId>org.springframework.boot</groupId>
44.                    <artifactId>spring-boot-maven-plugin</artifactId>
45.                </plugin>
46.                <plugin>
47.                    <groupId>org.jetbrains.kotlin</groupId>
48.                    <artifactId>kotlin-maven-plugin</artifactId>
49.                    <configuration>
50.                        <args>
51.                            <arg>-Xjsr305=strict</arg>
52.                        </args>
53.                        <compilerPlugins>
54.                            <plugin>spring</plugin>
55.                            <plugin>jpa</plugin>
56.                        </compilerPlugins>
57.                    </configuration>
58.                    <dependencies>
59.                        <dependency>
60.                            <groupId>org.jetbrains.kotlin</groupId>
61.                            <artifactId>kotlin-maven-allopen</artifactId>
62.                            <version>${kotlin.version}</version>
63.                        </dependency>
64.                        <dependency>
65.                            <groupId>org.jetbrains.kotlin</groupId>
66.                            <artifactId>kotlin-maven-noarg</artifactId>
67.                            <version>${kotlin.version}</version>
68.                        </dependency>
69.                    </dependencies>
70.                </plugin>
71.            </plugins>
72.        </build>
73.    </project>
```

application.yml 文件的内容如下：

```yaml
1.  spring:
2.    application:
3.      name: configserver       #应用名
4.    cloud:
5.      config:
6.        server:
7.          git:
8.            uri: https://github.com/dutyk/kotlinmicroservice   #Git仓库地址
9.            username:      #Git用户名
10.           password:      #Git密码
11.           default-label: master #配置文件分支
12.           search-paths: config  #配置文件所在根目录
13. server:
14.   port: 8060   #应用端口号
```

在 GitHub 上上传一个配置文件：springcloudconfig-dev.yml，定义如下：

```yaml
1. data:
2.   env: dev
3.   user:
4.     username: user_dev01
5.     password: password_dev
```

ConfigApplication.kt 定义了一个启动类，启动一个配置中心：

```kotlin
1. // 开启 config server 注解
2. @SpringBootApplication
3. @EnableConfigServer
4. class ConfigApplication
5. // 启动函数
6. fun main(args: Array<String>) {
7.     runApplication<ConfigApplication>(*args)
8. }
```

新建一个 Maven 子工程：chapter06-springcloud-config-client，这是一个客户端，从 config server 获取配置。pom 文件如下：

```xml
1.  <?xml version="1.0" encoding="UTF-8"?>
2.  <project xmlns="http://maven.apache.org/POM/4.0.0"
3.           xmlns:xsi="http://www.w3.org/2001/XMLSchema-instance"
4.           xsi:schemaLocation="http://maven.apache.org/POM/4.0.0 http://maven.apache.org/xsd/maven-4.0.0.xsd">
5.      <parent>
6.          <artifactId>kotlinspringboot</artifactId>
7.          <groupId>io.kang.kotlinspringboot</groupId>
8.          <version>0.0.1-SNAPSHOT</version>
9.      </parent>
10.     <modelVersion>4.0.0</modelVersion>
11.     <!-- 子工程名 -->
12.     <artifactId>chapter06-springcloud-config-client</artifactId>
13.     <dependencies>
14.         <!-- Spring Cloud Config 依赖包 -->
15.         <dependency>
16.             <groupId>org.springframework.cloud</groupId>
17.             <artifactId>spring-cloud-starter-config</artifactId>
18.             <version>2.2.1.RELEASE</version>
19.         </dependency>
20.         <!-- Spring Boot Web 依赖包 -->
21.         <dependency>
22.             <groupId>org.springframework.boot</groupId>
23.             <artifactId>spring-boot-starter-web</artifactId>
24.             <version>2.2.1.RELEASE</version>
25.         </dependency>
26.         <!-- Spring Boot Actuator 依赖包 -->
27.         <dependency>
28.             <groupId>org.springframework.boot</groupId>
29.             <artifactId>spring-boot-starter-actuator</artifactId>
30.             <version>2.2.1.RELEASE</version>
31.         </dependency>
32.         <dependency>
33.             <groupId>com.fasterxml.jackson.module</groupId>
34.             <artifactId>jackson-module-kotlin</artifactId>
35.         </dependency>
```

```xml
36.    <dependency>
37.        <groupId>org.jetbrains.kotlin</groupId>
38.        <artifactId>kotlin-reflect</artifactId>
39.    </dependency>
40.    <dependency>
41.        <groupId>org.jetbrains.kotlin</groupId>
42.        <artifactId>kotlin-stdlib-jdk8</artifactId>
43.    </dependency>
44.    <dependency>
45.        <groupId>org.jetbrains.kotlinx</groupId>
46.        <artifactId>kotlinx-coroutines-core</artifactId>
47.        <version>1.3.2</version>
48.    </dependency>
49.  </dependencies>
50.
51.  <build>
52.    <sourceDirectory>${project.basedir}/src/main/kotlin</sourceDirectory>
53.    <testSourceDirectory>${project.basedir}/src/test/kotlin</testSourceDirectory>
54.    <plugins>
55.      <plugin>
56.        <groupId>org.springframework.boot</groupId>
57.        <artifactId>spring-boot-maven-plugin</artifactId>
58.      </plugin>
59.      <plugin>
60.        <groupId>org.jetbrains.kotlin</groupId>
61.        <artifactId>kotlin-maven-plugin</artifactId>
62.        <configuration>
63.          <args>
64.            <arg>-Xjsr305=strict</arg>
65.          </args>
66.          <compilerPlugins>
67.            <plugin>spring</plugin>
68.            <plugin>jpa</plugin>
69.          </compilerPlugins>
70.        </configuration>
71.        <dependencies>
72.          <dependency>
```

```xml
73.                    <groupId>org.jetbrains.kotlin</groupId>
74.                    <artifactId>kotlin-maven-allopen</artifactId>
75.                    <version>${kotlin.version}</version>
76.                </dependency>
77.                <dependency>
78.                    <groupId>org.jetbrains.kotlin</groupId>
79.                    <artifactId>kotlin-maven-noarg</artifactId>
80.                    <version>${kotlin.version}</version>
81.                </dependency>
82.            </dependencies>
83.        </plugin>
84.    </plugins>
85. </build>
86. </project>
```

application.yml 文件的内容如下:

```yml
1. management:
2.   endpoints:
3.     web:
4.       exposure:
5.         include: refresh,health,info   #暴露监控接口
```

同时，还定义了一个 bootstrap.yml：

```yml
1.  server:
2.    port: 8061   # 应用端口
3.  spring:
4.    application:
5.      name: springcloudconfig   # 应用名称
6.    cloud:
7.      config:
8.        label: master   # 获取 master 分支
9.        profile: dev   # 获取 application-dev.yml 文件配置
10.       uri: http://localhost:8060   # 配置中心地址
```

ClientApplication.kt 定义了一个启动类，启动一个 Spring Boot Web 应用：

```
1. @SpringBootApplication
```

```
2.  class ClientApplication
3.  // 启动函数
4.  fun main(args: Array<String>) {
5.      runApplication<ClientApplication>(*args)
6.  }
```

GitConfig.kt 定义了一些属性及相应的值，这些属性从配置中心获取：

```
1.  @Component
2.  // 开启刷新
3.  @RefreshScope
4.  // 和配置文件定义的值一一对应
5.  data class GitConfig(
6.          @Value("\${data.env}")
7.          val env: String,
8.          @Value("\${data.user.username}")
9.          val username: String,
10.         @Value("\${data.user.password}")
11.         val password: String
12. )
```

ConfigController.kt 定义了一个测试接口：

```
1.  @RestController
2.  class ConfigController {
3.      @Autowired
4.      lateinit var gitConfig: GitConfig
5.      // 测试接口，用于获取配置
6.      @GetMapping("/config")
7.      fun getConfig(): String {
8.          return gitConfig.toString()
9.      }
10. }
```

访问"/config"接口，可以获取到 env、username、password 这几个配置项。修改 env 的值，提交到 Git 仓库，然后调用"actuator/refresh"刷新配置，再访问"/config"，可获取最新的值。

6.2 Apollo 配置中心

Apollo 是携程研发的配置中心,能够集中管理应用程序在开发、测试、生产环境中的配置信息。Apollo 提供了一个统一界面集中管理配置,支持多环境、多数据中心配置及权限管理等特性。本节介绍使用 Apollo 作为 Kotlin 开发的微服务的配置中心的相关知识。

6.2.1 Apollo 介绍

Apollo 中的配置是独立于程序的只读变量,其伴随应用的整个生命周期。其中的配置可以有多种加载方式,常见的有程序内部的硬编码、配置文件、环境变量、启动参数、基于数据库等。配置需要治理,需要进行权限控制,以管理不同环境和集群的配置。Apollo 从 4 个维度管理 key-value 的配置。

- application(应用):每个应用都需要有唯一的身份标识 appId。
- environment(环境):环境默认是通过读取机器上的配置(server.properties 中的 env 属性)指定的,也支持运行时通过系统属性等指定。
- cluster(集群):集群默认是通过读取机器上的配置(server.properties 中的 idc 属性)指定的,也支持运行时通过系统属性指定)。
- namespace(命名空间):一个应用的不同配置的分组。可以简单地把 namespace 类比为文件,不同类型的配置存放在不同的文件中,可以直接读取,也可继承公共配置。同一份代码部署在不同的集群,可以有不同的配置,比如 Zookeeper 的地址等。通过命名空间可以很方便地支持多个不同应用共享同一份配置,同时还允许应用对共享的配置进行覆盖。

用户在 Apollo 中修改完配置并发布后,客户端能实时(1 秒)接收到最新的配置,并通知到应用程序。所有的配置发布都有版本概念,从而可以方便地支持配置的回滚。Apollo 还支持配置的灰度发布,比如发布后,只对部分应用实例生效,等观察一段时间没问题后再推给所有应用实例。Apollo 对应用和配置的管理有完善的权限管理机制,对配置的管理还分为编辑和发布两个环节,可减少人为造成的错误。

Apollo 目前唯一的外部依赖是 MySQL,所以部署非常简单,只要安装好 Java 和 MySQL 就可以让 Apollo 运行起来。Apollo 还提供了打包脚本,一键就可以生成所有需要的安装包,并且支持自定义运行时参数。

Apollo 的核心组件有如下几个。

ConfigService:提供配置获取接口,提供配置推送接口,服务于 Apollo 客户端。

AdminService：提供配置管理接口，提供配置修改发布接口，服务于管理界面 Portal。

Client：为应用获取配置，支持实时更新，通过 MetaServer 获取 ConfigService 的服务列表，使用客户端软负载 SLB 方式调用 ConfigService。

Portal：配置管理界面，通过 MetaServer 获取 AdminService 的服务列表，使用客户端软负载 SLB 方式调用 AdminService。

Eureka：用于服务发现和注册，ConfigService 和 AdminService 在 Eureka 注册实例并定期上报心跳，Eureka 和 ConfigService 在一起部署。

MetaServer：Portal 通过域名访问 MetaServer 获取 AdminService 的地址列表，Client 通过域名访问 MetaServer 获取 ConfigService 的地址列表。MetaServer 相当于一个 Eureka Proxy，MetaServer 和 ConfigService 在一起部署。

NginxLB：和域名系统配合，协助 Portal 访问 MetaServer 获取 AdminService 的地址列表；和域名系统配合，协助 Client 访问 MetaServer 获取 ConfigService 地址列表；和域名系统配合，协助用户访问 Portal 进行配置管理。

6.2.2　Kotlin 集成 Apollo

启动 Apollo 服务，在 Apollo 中添加如图 6.2 所示的配置。

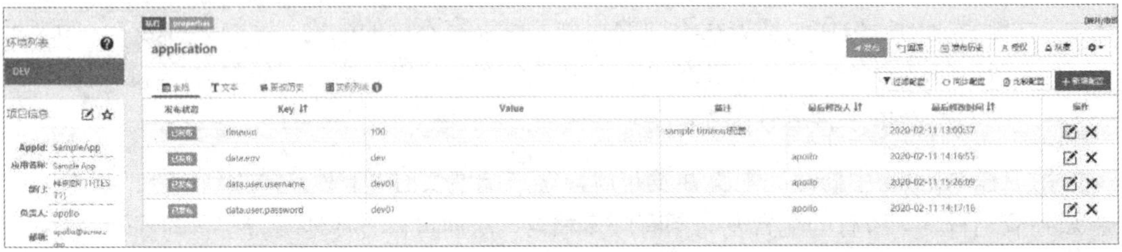

图6.2　Apollo的配置界面

新建一个 Maven 子工程：chapter06-apollo-config，从 Apollo 获取配置。pom 文件如下：

```
1.  <?xml version="1.0" encoding="UTF-8"?>
2.  <project xmlns="http://maven.apache.org/POM/4.0.0"
3.           xmlns:xsi="http://www.w3.org/2001/XMLSchema-instance"
4.           xsi:schemaLocation="http://maven.apache.org/POM/4.0.0
    http://maven.apache.org/xsd/maven-4.0.0.xsd">
5.      <parent>
```

```xml
6.     <artifactId>kotlinspringboot</artifactId>
7.     <groupId>io.kang.kotlinspringboot</groupId>
8.     <version>0.0.1-SNAPSHOT</version>
9.   </parent>
10.  <modelVersion>4.0.0</modelVersion>
11.  <!-- 子工程名 -->
12.  <artifactId>chapter06-apollo-config</artifactId>
13.  <dependencies>
14.    <!-- Apollo 客户端依赖包-->
15.    <dependency>
16.      <groupId>com.ctrip.framework.apollo</groupId>
17.      <artifactId>apollo-client</artifactId>
18.      <version>1.5.1</version>
19.    </dependency>
20.    <!-- Apollo 核心依赖包-->
21.    <dependency>
22.      <groupId>com.ctrip.framework.apollo</groupId>
23.      <artifactId>apollo-core</artifactId>
24.      <version>1.5.1</version>
25.    </dependency>
26.    <!-- Spring Boot Web 依赖包-->
27.    <dependency>
28.      <groupId>org.springframework.boot</groupId>
29.      <artifactId>spring-boot-starter-web</artifactId>
30.      <version>2.2.1.RELEASE</version>
31.    </dependency>
32.    <dependency>
33.      <groupId>com.fasterxml.jackson.module</groupId>
34.      <artifactId>jackson-module-kotlin</artifactId>
35.    </dependency>
36.    <dependency>
37.      <groupId>org.jetbrains.kotlin</groupId>
38.      <artifactId>kotlin-reflect</artifactId>
39.    </dependency>
40.    <dependency>
41.      <groupId>org.jetbrains.kotlin</groupId>
42.      <artifactId>kotlin-stdlib-jdk8</artifactId>
```

```xml
43.        </dependency>
44.        <dependency>
45.            <groupId>org.jetbrains.kotlinx</groupId>
46.            <artifactId>kotlinx-coroutines-core</artifactId>
47.            <version>1.3.2</version>
48.        </dependency>
49.    </dependencies>
50.    <build>
51.        <sourceDirectory>${project.basedir}/src/main/kotlin</sourceDirectory>
52.        <testSourceDirectory>${project.basedir}/src/test/kotlin</testSourceDirectory>
53.        <plugins>
54.            <plugin>
55.                <groupId>org.springframework.boot</groupId>
56.                <artifactId>spring-boot-maven-plugin</artifactId>
57.            </plugin>
58.            <plugin>
59.                <groupId>org.jetbrains.kotlin</groupId>
60.                <artifactId>kotlin-maven-plugin</artifactId>
61.                <configuration>
62.                    <args>
63.                        <arg>-Xjsr305=strict</arg>
64.                    </args>
65.                    <compilerPlugins>
66.                        <plugin>spring</plugin>
67.                        <plugin>jpa</plugin>
68.                    </compilerPlugins>
69.                </configuration>
70.                <dependencies>
71.                    <dependency>
72.                        <groupId>org.jetbrains.kotlin</groupId>
73.                        <artifactId>kotlin-maven-allopen</artifactId>
74.                        <version>${kotlin.version}</version>
75.                    </dependency>
76.                    <dependency>
77.                        <groupId>org.jetbrains.kotlin</groupId>
78.                        <artifactId>kotlin-maven-noarg</artifactId>
79.                        <version>${kotlin.version}</version>
```

```
80.                    </dependency>
81.                </dependencies>
82.            </plugin>
83.        </plugins>
84.    </build>
85. </project>
```

application.yml 文件的内容如下：

```
1. app:
2.   id: SampleApp  # app id 名
3. apollo:
4.   meta: http://127.0.0.1:8080  # Apollo 配置中心地址
5. server:
6.   port: 8062  # 应用端口号
```

ApolloApplication.kt 是一个启动类，启动一个 Spring Boot Web 应用：

```
1. // 开启 Apollo config 注解
2. @SpringBootApplication
3. @EnableApolloConfig
4. class ApolloApplication
5. // 启动函数
6. fun main(args: Array<String>) {
7.     System.setProperty("env", "dev")
8.     runApplication<ApolloApplication>(*args)
9. }
```

ApolloConfig.kt 定义了 Apollo 中的配置项并监听配置更新：

```
1. // 配置实体类
2. @Configuration
3. @ConfigurationProperties(prefix = "data")
4. class ApolloConfig {
5.     var env: String? = null
6.     var user: User? = null
7.     // 监听函数，当 username 值更新时触发该函数
8.     @ApolloConfigChangeListener
9.     fun configChangeHandlerUserName(configChangeEvent: ConfigChangeEvent)
   {
        if(configChangeEvent.isChanged("data.user.username")) {
```

```
10.              user?.username =
    configChangeEvent.getChange("data.user.username").newValue
11.              println("${user?.username} is change")
12.         }
13.     }
14. }
15. // User 实体类
16. class User{
17.     var username: String? = null
18.     var password: String? = null
19.     override fun toString(): String {
20.         return "${username},${password}"
21.     }
22. }
```

ApolloController.kt 定义了一个测试接口：

```
1. @RestController
2. class ApolloController {
3.     @Autowired
4.     lateinit var apolloConfig: ApolloConfig
5.     // 测试接口，测试从 Apollo 读取配置值
6.     @GetMapping("config")
7.     fun getApolloConfig(): String {
8.         return "${apolloConfig.env},${apolloConfig.user}"
9.     }
10. }
```

访问"config"接口，可以获取 env 和 user 的值。修改 env 的值，发布 Apollo 配置，再次访问"config"接口，可以获取最新的 env 值。

6.3 Nacos 配置中心

本节将介绍使用 Nacos 作为 Kotlin 开发的微服务的配置中心的相关知识。在 Nacos 的配置界面中新增了两个 Data ID：nacos-config-dev.yaml 和 nacos-config.yaml，并新增了一些配置项，如图 6.3 和图 6.4 所示。

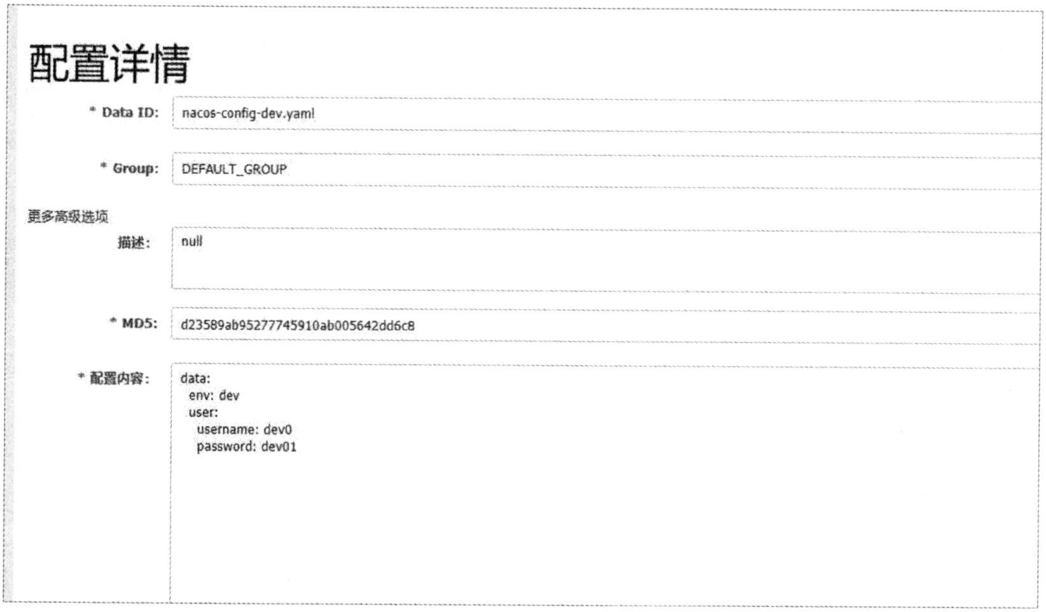

图6.3 nacos-config-dev.yaml配置详情

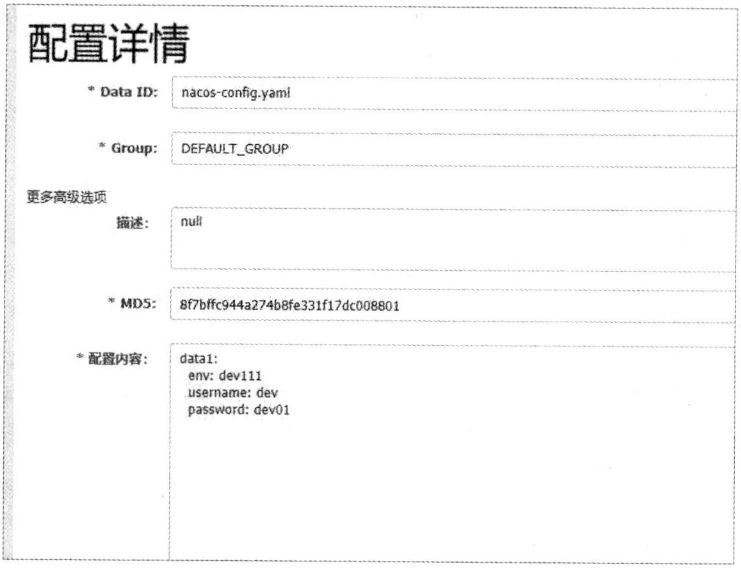

图6.4 nacos-config.yaml配置详情

新建一个 Maven 子工程：chapter06-nacos-config，这是一个客户端，可以从 Nacos 获取配置。pom 文件如下：

```xml
1.  <?xml version="1.0" encoding="UTF-8"?>
2.  <project xmlns="http://maven.apache.org/POM/4.0.0"
3.      xmlns:xsi="http://www.w3.org/2001/XMLSchema-instance"
4.      xsi:schemaLocation="http://maven.apache.org/POM/4.0.0 http://maven.apache.org/xsd/maven-4.0.0.xsd">
5.      <parent>
6.          <artifactId>kotlinspringboot</artifactId>
7.          <groupId>io.kang.kotlinspringboot</groupId>
8.          <version>0.0.1-SNAPSHOT</version>
9.      </parent>
10.     <modelVersion>4.0.0</modelVersion>
11.     <!-- 子工程名-->
12.     <artifactId>chapter06-nacos-config</artifactId>
13.     <dependencies>
14.         <!-- Spring Cloud Nacos 依赖包 -->
15.         <dependency>
16.             <groupId>com.alibaba.cloud</groupId>
17.             <artifactId>spring-cloud-starter-alibaba-nacos-config</artifactId>
18.             <version>2.2.0.RELEASE</version>
19.         </dependency>
20.         <!-- Spring Boot Web 依赖包 -->
21.         <dependency>
22.             <groupId>org.springframework.boot</groupId>
23.             <artifactId>spring-boot-starter-web</artifactId>
24.             <version>2.2.1.RELEASE</version>
25.         </dependency>
26.         <dependency>
27.             <groupId>com.fasterxml.jackson.module</groupId>
28.             <artifactId>jackson-module-kotlin</artifactId>
29.         </dependency>
30.         <dependency>
31.             <groupId>org.jetbrains.kotlin</groupId>
32.             <artifactId>kotlin-reflect</artifactId>
33.         </dependency>
34.         <dependency>
35.             <groupId>org.jetbrains.kotlin</groupId>
```

```xml
36.            <artifactId>kotlin-stdlib-jdk8</artifactId>
37.        </dependency>
38.        <dependency>
39.            <groupId>org.jetbrains.kotlinx</groupId>
40.            <artifactId>kotlinx-coroutines-core</artifactId>
41.            <version>1.3.2</version>
42.        </dependency>
43.    </dependencies>
44.    <build>
45.        <sourceDirectory>${project.basedir}/src/main/kotlin</sourceDirectory>
46.        <testSourceDirectory>${project.basedir}/src/test/kotlin</testSourceDirectory>
47.        <plugins>
48.            <plugin>
49.                <groupId>org.springframework.boot</groupId>
50.                <artifactId>spring-boot-maven-plugin</artifactId>
51.            </plugin>
52.            <plugin>
53.                <groupId>org.jetbrains.kotlin</groupId>
54.                <artifactId>kotlin-maven-plugin</artifactId>
55.                <configuration>
56.                    <args>
57.                        <arg>-Xjsr305=strict</arg>
58.                    </args>
59.                    <compilerPlugins>
60.                        <plugin>spring</plugin>
61.                        <plugin>jpa</plugin>
62.                    </compilerPlugins>
63.                </configuration>
64.                <dependencies>
65.                    <dependency>
66.                        <groupId>org.jetbrains.kotlin</groupId>
67.                        <artifactId>kotlin-maven-allopen</artifactId>
68.                        <version>${kotlin.version}</version>
69.                    </dependency>
70.                    <dependency>
71.                        <groupId>org.jetbrains.kotlin</groupId>
72.                        <artifactId>kotlin-maven-noarg</artifactId>
```

```
73.                    <version>${kotlin.version}</version>
74.                </dependency>
75.            </dependencies>
76.          </plugin>
77.      </plugins>
78.   </build>
79. </project>
```

bootstrap.yml 文件的内容如下所示，其中指定了配置中心的地址和配置文件的后缀。Nacos 的 data-id 名称由 spring.application.name 和 spring.profiles.active 组成，后缀是 yaml。此外，还可以用 extension-configs 指定 data-id：

```
1.  spring:
2.    cloud:
3.      nacos:
4.        config:
5.          server-addr: 127.0.0.1:8848  # Nacos 配置中心的地址
6.          file-extension: yaml  # 配置文件采用 yaml 格式
7.          extension-configs:
8.            -
9.              data-id: nacos-config.yaml  # data-id 定义
10.             group: default  # group 定义
11.             refresh: true  # 配置自动刷新
12.   profiles:
13.     active: dev  # 读取 dev 结尾的配置文件
14.   application:
15.     name: nacos-config  # 应用名
16. server:
17.   port: 8063  # 应用端口号
```

NacosApplication.kt 是一个启动类，启动了一个 Spring Boot Web 应用：

```
1. @SpringBootApplication
2. class NacosApplication
3. // 启动函数
4. fun main(args: Array<String>) {
5.     runApplication<NacosApplication>(*args)
6. }
```

NacosConfig.kt 定义了 nacos-config-dev.yaml 的配置项。当更新 Nacos 配置时，使用 @ConfigurationProperties 注解，NacosConfig 这个类的属性值也会更新：

```kotlin
1.  // 实体类，对应 nacos 的配置项
2.  @Component
3.  @ConfigurationProperties(prefix = "data")
4.  class NacosConfig {
5.      var env: String? = null
6.      var user: User? = null
7.  }
8.  // User 实体类
9.  class User {
10.     var username: String? = null
11.     var password: String? = null
12.     override fun toString(): String {
13.         return "${username},${password}"
14.     }
15. }
```

NacosConfig1.kt 定义了 nacos-config.yaml 的配置项。使用@Value 注解，Nacos 的配置更新，NacosConfig1 的属性值不会更新，重启应用才会更新：

```kotlin
1.  // 实体类，对应 Nacos 配置项
2.  @Configuration
3.  class NacosConfig1 {
4.      @Value(value = "\${data1.env}")
5.      var env: String? = null
6.      @Value(value = "\${data1.username}")
7.      var username: String? = null
8.      @Value(value = "\${data1.password}")
9.      var password: String? = null
10. }
```

NacosController.kt 定义了两个测试接口：

```kotlin
1.  @RestController
2.  class NacosController {
3.      @Autowired
4.      lateinit var nacosConfig: NacosConfig
5.      @Autowired
```

```
6.    lateinit var nacosConfig1: NacosConfig1
7.    // 测试接口，读取 Nacos 配置项
8.    @GetMapping("config")
9.    fun getNacosConfig(): String {
10.       return "${nacosConfig.env}-${nacosConfig.user}"
11.   }
12.   // 测试接口，读取 Nacos 配置项
13.   @GetMapping("config1")
14.   fun getNacosConfig1(): String {
15.       return "${nacosConfig1.env}-${nacosConfig1.username}-${nacosConfig1.password}"
16.   }
17. }
```

访问 "config" 接口可以获取 nacos-config-dev.yaml 的配置项。更新 Nacos 的配置，并发布，可以自动更新配置，再次访问 "config" 接口，可以获取最新的配置。访问 "config1" 可以获取 nacos-config.yaml 对应的配置。

6.4 Consul 配置中心

本节介绍使用 Consul 作为 Kotlin 开发的微服务的配置中心的相关知识。在 Consul 中配置如图 6.5 所示的属性。

图6.5 Consul配置界面

新建一个 Maven 子工程：chapter06-consul-config，它从 Consul 读取配置，pom 文件如下：

```xml
1.  <?xml version="1.0" encoding="UTF-8"?>
2.  <project xmlns="http://maven.apache.org/POM/4.0.0"
3.           xmlns:xsi="http://www.w3.org/2001/XMLSchema-instance"
4.           xsi:schemaLocation="http://maven.apache.org/POM/4.0.0 http://maven.apache.org/xsd/maven-4.0.0.xsd">
5.      <parent>
6.          <artifactId>kotlinspringboot</artifactId>
7.          <groupId>io.kang.kotlinspringboot</groupId>
8.          <version>0.0.1-SNAPSHOT</version>
9.      </parent>
10.     <modelVersion>4.0.0</modelVersion>
11.     <!-- 子工程名-->
12.     <artifactId>chapter06-consul-config</artifactId>
13.     <dependencies>
14.         <!-- Spring Cloud Consul Config 依赖包-->
15.         <dependency>
16.             <groupId>org.springframework.cloud</groupId>
17.             <artifactId>spring-cloud-starter-consul-config</artifactId>
18.             <version>2.2.1.RELEASE</version>
19.         </dependency>
20.         <!-- Spring Boot Web 依赖包-->
21.         <dependency>
22.             <groupId>org.springframework.boot</groupId>
23.             <artifactId>spring-boot-starter-web</artifactId>
24.             <version>2.2.1.RELEASE</version>
25.         </dependency>
26.         <!-- Spring Boot Actuator 依赖包-->
27.         <dependency>
28.             <groupId>org.springframework.boot</groupId>
29.             <artifactId>spring-boot-starter-actuator</artifactId>
30.             <version>2.2.1.RELEASE</version>
31.         </dependency>
32.         <dependency>
33.             <groupId>com.fasterxml.jackson.module</groupId>
```

```xml
34.        <artifactId>jackson-module-kotlin</artifactId>
35.    </dependency>
36.    <dependency>
37.        <groupId>org.jetbrains.kotlin</groupId>
38.        <artifactId>kotlin-reflect</artifactId>
39.    </dependency>
40.    <dependency>
41.        <groupId>org.jetbrains.kotlin</groupId>
42.        <artifactId>kotlin-stdlib-jdk8</artifactId>
43.    </dependency>
44.    <dependency>
45.        <groupId>org.jetbrains.kotlinx</groupId>
46.        <artifactId>kotlinx-coroutines-core</artifactId>
47.        <version>1.3.2</version>
48.    </dependency>
49. </dependencies>
50. <build>
51.     <sourceDirectory>${project.basedir}/src/main/kotlin</sourceDirectory>
52.     <testSourceDirectory>${project.basedir}/src/test/kotlin</testSourceDirectory>
53.     <plugins>
54.         <plugin>
55.             <groupId>org.springframework.boot</groupId>
56.             <artifactId>spring-boot-maven-plugin</artifactId>
57.         </plugin>
58.         <plugin>
59.             <groupId>org.jetbrains.kotlin</groupId>
60.             <artifactId>kotlin-maven-plugin</artifactId>
61.             <configuration>
62.                 <args>
63.                     <arg>-Xjsr305=strict</arg>
64.                 </args>
65.                 <compilerPlugins>
66.                     <plugin>spring</plugin>
67.                     <plugin>jpa</plugin>
68.                 </compilerPlugins>
69.             </configuration>
70.             <dependencies>
```

```xml
71.            <dependency>
72.                <groupId>org.jetbrains.kotlin</groupId>
73.                <artifactId>kotlin-maven-allopen</artifactId>
74.                <version>${kotlin.version}</version>
75.            </dependency>
76.            <dependency>
77.                <groupId>org.jetbrains.kotlin</groupId>
78.                <artifactId>kotlin-maven-noarg</artifactId>
79.                <version>${kotlin.version}</version>
80.            </dependency>
81.            </dependencies>
82.            </plugin>
83.        </plugins>
84.    </build>
85. </project>
```

bootstrap.yml 文件的内容如下所示，其中指定了配置文件的格式、目录等信息：

```yaml
1.  spring:
2.    application:
3.      name: consul-config  # 应用名
4.    cloud:
5.      consul:
6.        host: 127.0.0.1  # Consul 配置中心地址
7.        port: 8500  # Consul 配置中心端口
8.        config:
9.          prefix: config  # 配置文件前缀
10.         enabled: true  # 是否生效
11.         format: yaml  # 配置文件格式
12.         data-key: data  # 配置项前缀
13.   profiles:
14.     active: dev  # 使用 dev 结尾的配置文件
15. server:
16.   port: 8065  # 应用端口号
```

ConsulConfigApplication.kt 定义了一个启动类：

```
1. @SpringBootApplication
```

```
2.  class ConsulConfigApplication
3.  // 启动函数
4.  fun main(args: Array<String>) {
5.      runApplication<ConsulConfigApplication>(*args)
6.  }
```

ConsulConfig.kt 定义了 Consul 中的配置项。使用@ConfigurationProperties 注解将 ConsulConfig 定义的值和 Consul 中的配置对应起来:

```
1.  // 配置实体类,对应 Consul 定义的配置
2.  @Component
3.  @ConfigurationProperties(prefix = "data")
4.  class ConsulConfig {
5.      var env: String? = null
6.      var user: User? = null
7.  }
8.  // User 实体类
9.  class User{
10.     var username: String? = null
11.     var password: String? = null
12.     override fun toString(): String {
13.         return "${username},${password}"
14.     }
15. }
```

ConsulController.kt 定义了一个测试接口:

```
1.  @RestController
2.  class ConsulController {
3.      @Autowired
4.      lateinit var consulConfig: ConsulConfig
5.      // 测试接口,从 Consul 读取配置
6.      @GetMapping("config")
7.      fun getConsulConfig(): String {
8.          return "${consulConfig.env}, ${consulConfig.user}"
9.      }
10. }
```

访问"config"可以从 Consul 获取配置值。在 Consul 中修改配置值，再次访问"config"接口，可以获取最新的配置。

6.5 小结

本章介绍了 Kotlin 集成 Spring Cloud Config、Apollo、Nacos、Consul 这四种配置中心的相关知识。对微服务系统来说，一个系统可以划分为很多个微服务，每个微服务都有自己的配置。配置中心可帮助高效管理微服务的配置，配置的热加载机制使得不需要重启应用，配置就可生效。

第 7 章
Kotlin 应用于微服务网关

网关是微服务系统对外提供服务的窗口，通过网关，可将请求分发到不同的子系统。本章将介绍 Zuul 和 Spring Cloud Gateway，并通过示例介绍 Kotlin 开发微服务网关的方法。

7.1 Kotlin 集成 Zuul

Zuul 是 Netflix 公司开源的一个 API Gateway 组件，是所有从设备和 Web 站点到 Netflix 流媒体应用程序后端请求的前门。本节将介绍 Zuul 的架构和原理，以及使用 Kotlin 集成 Zuul 进行微服务网关开发的相关知识。

7.1.1 Zuul 介绍

作为一个边缘服务应用程序，Zuul 支持动态路由、监视，并为请求提供弹性和安全性。它还可以根据需要将请求路由到多个 Amazon 自动伸缩组。Zuul 使用了一系列不同类型的过滤器，使我们能够快速灵活地将功能应用到边缘服务中。这些过滤器可帮助我们执行以下功能。

- **身份验证和安全性**：满足每个资源的身份验证需求并拒绝不满足这些需求的请求。
- **洞察和监控**：在边缘跟踪有意义的数据和统计数据，以便提供准确的生产视图。

- **动态路由**：将外部的请求转发到具体的微服务实例上，这是实现外部访问统一入口的基础。
- **压力测试**：逐步增加集群的流量，以评估性能。
- **减少负载**：为每种类型的请求分配容量，并删除超过限制的请求。
- **静态响应处理**：直接在边缘构建一些响应，而不是将它们转发到内部集群。
- **多区域弹性**：跨 AWS 区域路由请求，以使 ELB 使用多样化。

Zuul 目前有两个版本。Zuul1 设计比较简单，本质上就是一个同步 Servlet，采用多线程阻塞模型，同步阻塞模式的编程模型比较简单，开发调试运维也比较简单。但是线程上下文切换开销大，连接数量受限制，延迟阻塞会耗尽线程资源。Zuul2 使用 Netty 实现异步非阻塞编程模型，线程开销小，连接数量易于扩展，但是编程模型复杂，开发调试运维复杂。本书使用 Zuul1 版本进行介绍。

Zuul1 过滤器的原理如图 7.1 所示，过滤器分为四种：pre、post、routing 及 error。

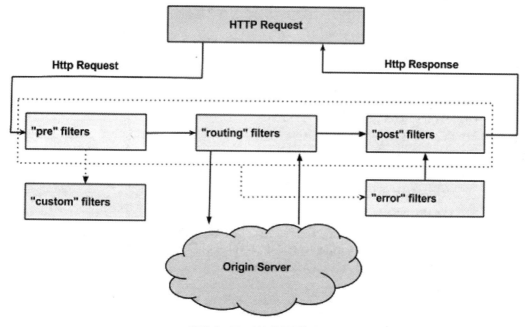

图7.1 Zuul1过滤器的原理

Zuul2 基于 Netty，它首先运行预过滤器（入站过滤器），然后使用 Netty 客户机代理请求，最后在运行后过滤器（出站过滤器）后返回响应。

Zuul 和 Eureka 进行整合，将 Zuul 自身注册为 Eureka 服务治理下的应用，Eureka 注册的其他应用可以通过 Zuul 跳转后访问。Spring Cloud 对 Zuul 进行了整合与增强，Zuul 默认使用的 HTTP 客户端是 Apache HTTPClient，也可以使用 RESTClient 或 OkHttp3 中的 OkHttpClient。Zuul 默认和 Ribbon 结合实现负载均衡的功能。Zuul 的路由配置信息可以放在 Spring Cloud Config 中，实现动态调整。

7.1.2　Kotlin 集成 Zuul

新建 Maven 子工程 "chapter07-zuul"，这是一个基于 Zuul 的网关微服务。pom 文件如下：

```xml
1.  <?xml version="1.0" encoding="UTF-8"?>
2.  <project xmlns="http://maven.apache.org/POM/4.0.0"
3.      xmlns:xsi="http://www.w3.org/2001/XMLSchema-instance"
4.      xsi:schemaLocation="http://maven.apache.org/POM/4.0.0
    http://maven.apache.org/xsd/maven-4.0.0.xsd">
5.      <parent>
6.          <artifactId>kotlinspringboot</artifactId>
7.          <groupId>io.kang.kotlinspringboot</groupId>
8.          <version>0.0.1-SNAPSHOT</version>
9.      </parent>
10.     <modelVersion>4.0.0</modelVersion>
11.     <!-- 子工程名-->
12.     <artifactId>chapter07-zuul</artifactId>
13.     <dependencies>
14.         <!-- Spring Cloud Zuul 依赖包-->
15.         <dependency>
16.             <groupId>org.springframework.cloud</groupId>
17.             <artifactId>spring-cloud-starter-netflix-zuul</artifactId>
18.             <version>2.2.1.RELEASE</version>
19.         </dependency>
20.         <!-- Spring Cloud Config 依赖包-->
21.         <dependency>
22.             <groupId>org.springframework.cloud</groupId>
23.             <artifactId>spring-cloud-starter-config</artifactId>
24.             <version>2.2.1.RELEASE</version>
25.         </dependency>
26.         <!-- Spring Cloud Eureka 客户端依赖包-->
```

```xml
27.    <dependency>
28.        <groupId>org.springframework.cloud</groupId>
29.        <artifactId>spring-cloud-starter-netflix-eureka-client</artifactId>
30.        <version>2.2.1.RELEASE</version>
31.    </dependency>
32.    <!-- Spring Boot Web 依赖包-->
33.    <dependency>
34.        <groupId>org.springframework.boot</groupId>
35.        <artifactId>spring-boot-starter-web</artifactId>
36.        <version>2.2.1.RELEASE</version>
37.    </dependency>
38.    <!-- Spring Boot Actuator 依赖包-->
39.    <dependency>
40.        <groupId>org.springframework.boot</groupId>
41.        <artifactId>spring-boot-starter-actuator</artifactId>
42.        <version>2.2.1.RELEASE</version>
43.    </dependency>
44.    <dependency>
45.        <groupId>com.fasterxml.jackson.module</groupId>
46.        <artifactId>jackson-module-kotlin</artifactId>
47.    </dependency>
48.    <dependency>
49.        <groupId>org.jetbrains.kotlin</groupId>
50.        <artifactId>kotlin-reflect</artifactId>
51.    </dependency>
52.    <dependency>
53.        <groupId>org.jetbrains.kotlin</groupId>
54.        <artifactId>kotlin-stdlib-jdk8</artifactId>
55.    </dependency>
56.    <dependency>
57.        <groupId>org.jetbrains.kotlinx</groupId>
58.        <artifactId>kotlinx-coroutines-core</artifactId>
59.        <version>1.3.2</version>
60.    </dependency>
61. </dependencies>
62. <build>
```

```xml
63.         <sourceDirectory>${project.basedir}/src/main/kotlin</sourceDirectory>
64.         <testSourceDirectory>${project.basedir}/src/test/kotlin</testSourceDirectory>
65.         <plugins>
66.             <plugin>
67.                 <groupId>org.springframework.boot</groupId>
68.                 <artifactId>spring-boot-maven-plugin</artifactId>
69.             </plugin>
70.             <plugin>
71.                 <groupId>org.jetbrains.kotlin</groupId>
72.                 <artifactId>kotlin-maven-plugin</artifactId>
73.                 <configuration>
74.                     <args>
75.                         <arg>-Xjsr305=strict</arg>
76.                     </args>
77.                     <compilerPlugins>
78.                         <plugin>spring</plugin>
79.                         <plugin>jpa</plugin>
80.                     </compilerPlugins>
81.                 </configuration>
82.                 <dependencies>
83.                     <dependency>
84.                         <groupId>org.jetbrains.kotlin</groupId>
85.                         <artifactId>kotlin-maven-allopen</artifactId>
86.                         <version>${kotlin.version}</version>
87.                     </dependency>
88.                     <dependency>
89.                         <groupId>org.jetbrains.kotlin</groupId>
90.                         <artifactId>kotlin-maven-noarg</artifactId>
91.                         <version>${kotlin.version}</version>
92.                     </dependency>
93.                 </dependencies>
94.             </plugin>
95.         </plugins>
96.     </build>
97. </project>
```

application.yml 文件的内容如下所示，其中定义了路由信息，接口 "/provide/*" 的调用被发送给 provider-server 这个微服务，对微服务 consumer-feign 不做代理：

```yaml
eureka:
  client:
    service-url:
      defaultZone: http://localhost:8761/eureka/  #Eureka 注册中心地址
server:
  port: 8070   # 应用端口号
zuul:
  routes:
    provide-service-url:
      path: /provide/*  # 请求地址
      service-id: provider-server  # 被路由的服务 id
  ignored-services: consumer-feign  # 不对该服务进行路由代理
ribbon:
  eureka:
    enabled: true  # 使用 Ribbon 进行负载均衡
spring:
  application:
    name: zuul-app  # 应用名称
```

ZuulApplication.kt 是启动类，服务启动后会注册到 Eureka：

```kotlin
// 开启服务注册注解，Zuul 代理注解
@EnableDiscoveryClient
@EnableZuulProxy
@SpringBootApplication
class ZuulApplication
// 启动函数
fun main(args: Array<String>) {
    runApplication<ZuulApplication>(*args)
}
```

首先启动 Eureka 注册中心——chapter05-eureka，然后启动两个服务——chapter05-eureka-provider、chapter05-eureka-consumer，它们在 Eureka 中的 service-id 是 provider-server、consumer-feign，再启动网关服务 chapter07-zuul。

调用"provide/provide"会转发到 provider-server 的"provide"接口。如果 provider-server 有多个实例，通过设置 ribbon.eureka.enabled=true，Ribbon 会根据发现机制来获取配置服务名对应的实例清单，进行负载均衡。

Zuul 在注册到 Eureka 服务中心之后，它会为 Eureka 中的每个服务都创建一个默认的路由规则，默认规则的 path 会使用 service-id 配置的服务名作为请求前缀。在 application.yml 中将 ignored-services 注释掉，保留 consumer-feign 这个微服务。调用"consumer-feign/feignProvide"可以将请求转发到 consumer-feign 的"feignProvide"接口。

Zuul 路由的 path 属性通常需要通配符，通配符匹配规则如下。

- ?：匹配单个字符，如"/feign/?"。
- *：匹配任意数量字符，但不支持多级目录，如"/feign/*"。
- **：匹配任意数量字符，支持多级目录，如"/feign/**"。

AccessFilter.kt 定义了一个过滤器，校验请求是否包含"accessToken"参数。如果没有"accessToken"，返回 401 码值，如图 7.2 所示。

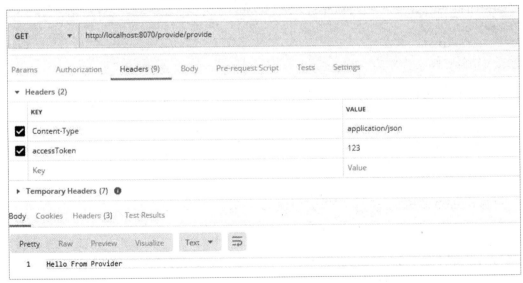

图7.2 /provide/provide接口调用结果

```
1.  @Component
2.  class AccessFilter: ZuulFilter() {
3.      // 是否过滤
```

```kotlin
4.      override fun shouldFilter(): Boolean {
5.          return true
6.      }
7.      // 过滤器类型
8.      override fun filterType(): String {
9.          return "pre"
10.     }
11.     // 拦截 header 的 accessToken，不能为空
12.     override fun run(): Any? {
13.         val ctx = RequestContext.getCurrentContext()
14.         val request = ctx.request
15.         println("进入访问过滤器，访问的 url:${request.requestURL}，访问的方法：${request.method}")
16.         val accessToken = request.getHeader("accessToken")
17.         if(accessToken == null || accessToken.isEmpty()) {
18.             ctx.setSendZuulResponse(false)
19.             ctx.responseStatusCode = 401
20.             return null
21.         }
22.         return null
23.     }
24.     // 过滤的优先级
25.     override fun filterOrder(): Int {
26.         return 0
27.     }
28. }
```

RateLimiterFilter.kt 定义了一个过滤器，校验接口的访问频率。接口访问频率过高，会返回异常，提示调用过多。用 postman 对 "provide/provide" 进行并发调用，并发数为 5，结果如图 7.3 所示。

图7.3　provide/provide并发测试结果

```
1.  @Component
2.  class RateLimiterFilter: ZuulFilter() {
3.      // 初始化限流器
4.      val rateLimiter = RateLimiter.create(2.0)
5.      // 过滤器类型
6.      override fun filterType(): String {
7.          return PRE_TYPE
8.      }
9.      // 过滤器优先级
10.     override fun filterOrder(): Int {
11.         return 1
12.     }
13.     // 对请求进行限流
14.     override fun run(): Any? {
15.         val ctx = RequestContext.getCurrentContext()
16.         if(!rateLimiter.tryAcquire()) {
17.             ctx.setSendZuulResponse(false)
18.             ctx.responseStatusCode = HttpStatus.TOO_MANY_REQUESTS.value()
19.         }
20.         return null
21.     }
22.     // 只对/provide/limit 接口执行该过滤器
```

```
23.    override fun shouldFilter(): Boolean {
24.        val ctx = RequestContext.getCurrentContext()
25.        val request = ctx.request
26.      if("/provide/limit" == request.requestURI){
27.            return true
28.        }
29.        return false
30.    }
31. }
```

ApiFallbackProvider.kt 定义了一个熔断器。如果其中一个服务"挂"掉了,那么请求就会进行漫长的超时等待,最终会返回失败,这甚至会影响整个服务链。熔断器可以及时处理"挂"掉的服务,及时给用户发送响应信息:

```
1. @Component
2. class ApiFallbackProvider: FallbackProvider {
3.     // 对所有接口执行熔断处理
4.     override fun getRoute(): String {
5.         return "*"
6.     }
7.     // 熔断处理函数
8.     override fun fallbackResponse(route: String?, cause: Throwable?): ClientHttpResponse {
9.         var message = ""
10.        if (cause is HystrixTimeoutException) {
11.            message = "Timeout"
12.        } else {
13.            message = "Service exception"
14.        }
15.        return fallbackResponse(message)
16.    }
17.    // 封装 ClienthttpResponse
18.    fun fallbackResponse(message: String): ClientHttpResponse {
19.        return object : ClientHttpResponse {
20.            // 返回状态码
21.            @Throws(IOException::class)
```

```kotlin
22.        override fun getStatusCode(): HttpStatus {
23.            return HttpStatus.OK
24.        }
25.        // 返回状态码
26.        @Throws(IOException::class)
27.        override fun getRawStatusCode(): Int {
28.            return 200
29.        }
30.        // 返回状态描述
31.        @Throws(IOException::class)
32.        override fun getStatusText(): String {
33.            return "OK"
34.        }
35.        override fun close() {
36.        }
37.        // 返回的 body
38.        @Throws(IOException::class)
39.        override fun getBody(): InputStream {
40.            val bodyText = String.format("{\"code\": 999,\"message\": \"Service unavailable:%s\"}", message)
41.            return ByteArrayInputStream(bodyText.toByteArray())
42.        }
43.        // HTTP 请求头
44.        override fun getHeaders(): HttpHeaders {
45.            val headers = HttpHeaders()
46.            headers.setContentType(MediaType.APPLICATION_JSON)
47.            return headers
48.        }
49.    }
50. }
51. }
```

将服务"provider-server"停掉，再次访问接口"provide/provide"会返回默认值，如图 7.4 所示。

第 7 章 Kotlin 应用于微服务网关

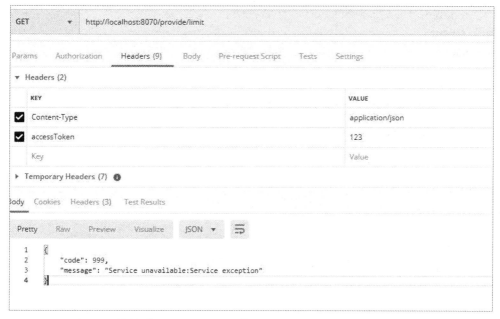

图7.4 停掉"provider-server", provide/provide接口的调用结果

为了实现动态路由，可以将 Zuul 的路由配置放在配置中心，如 Spring Cloud Config。在 Spring Cloud Config 中新建一个文件：zuulapp-dev.yml，上传到 Git 仓库：

```
1. zuul:
2.   routes:
3.     provide-service-url:
4.       path: /provide/* # 请求地址
5.       service-id: provider-server # Zuul 代理的服务 id
6.   ignored-services: consumer-feign # Zuul 屏蔽的服务
```

chapter07-zuul 子工程的 application.yml 文件调整如下：

```
1. eureka:
2.   client:
3.     service-url:
4.       defaultZone: http://localhost:8761/eureka/ # Eureka 注册中心地址
5. server:
6.   port: 8070 # 服务端口号
7. ribbon:
8.   eureka:
```

```yaml
9.      enabled: true    # 使用 Ribbon 进行负载均衡
10. management:
11.   endpoints:
12.     web:
13.       exposure:
14.         include: refresh,health,info   # 对外暴露的监控接口
```

新增一个配置文件 bootstrap.yml：

```yaml
1. spring:
2.   cloud:
3.     config:
4.       label: master      # 配置文件所在分支
5.       profile: dev       # dev 结尾的配置文件
6.       uri: http://localhost:8060   # 注册中心地址
7.   application:
8.     name: zuulapp         # 应用名
```

ZuulProperty.kt 定义了 Zuul 的相关配置：

```kotlin
1. // 配置实体类，对应 zuulapp-dev.yml 中定义的属性
2. @Configuration
3. class ZuulProperty {
4.     @Bean
5.     @ConfigurationProperties("zuul")
6.     @RefreshScope
7.     @Primary
8.     fun zuulProperties(): ZuulProperties {
9.         return ZuulProperties()
10.    }
11. }
```

依次启动 chapter05-eureka、chapter05-eureka-provider、chapter05-eureka-consumer、chapter06-springcloud-config 和 chapter07-zuul，然后访问"consumer-feign/feignProvide"接口。由于 zuulconfig-dev.yml 配置中屏蔽了"consumer-feign"这个服务，所以调用不能转发到"feignProvide"接口。修改 zuulconfig-dev.yml，注释掉"ignored-services"这个配置项，并提交到远程仓库。然后在 chapter07-zuul 工程中调用"actuator/refresh"刷新配置，再次访问"consumer-feign/feignProvide"接口，可以把服务转发到"feignProvide"接口。

7.2　Kotlin 集成 Spring Cloud Gateway

Spring Cloud Gateway 是 Spring 官方基于 Spring 5.0、Spring Boot 2.0 和 Project Reactor 等技术开发的网关，旨在为微服务架构提供一种简单有效、统一的 API 路由管理方式，并统一访问接口。本节介绍 Kotlin 集成 Spring Cloud Gateway 进行微服务网关开发的相关知识。

7.2.1　Spring Cloud Gateway 介绍

Spring Cloud Gateway 作为 Spring Cloud 生态系统中的网关，目标是替代 Netflix Zuul，其不仅提供了统一的路由方式，并且基于 Filter 链的方式提供了网关应具备的基本功能，例如，安全、监控/埋点和限流等。它是基于 Nttey 的响应式开发模式，使用的 Web 框架是 WebFlux。WebFlux 是一个典型的非阻塞的异步的框架，它的核心是基于 Reactor 的相关 API 实现的，具有非阻塞、函数式编程特点。

Spring Cloud Gateway 的工作流程如图 7.5 所示。

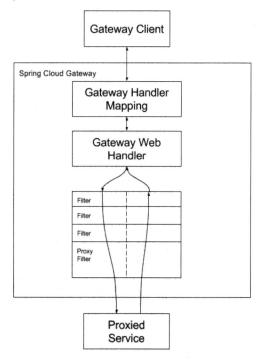

图7.5　Spring Cloud Gateway工作流程图

客户端向 Spring Cloud Gateway 发出请求，然后在 Gateway Handler Mapping 中找到与

请求相匹配的路由，将其发送到 Gateway Web Handler。Handler 再通过指定的过滤器链将请求发送到实际的服务执行业务逻辑，然后返回。过滤器之间用虚线分开是因为过滤器可能会在发送代理请求之前或之后执行业务逻辑。

路由（Route）是 Gateway 的基本构建模块。它由 ID、目标 URI、断言集合和过滤器集合组成。如果聚合断言结果为真，则匹配到该路由。

断言（Predicate）是一个 Java 8 Function Predicate。输入类型是 Spring Framework ServerWebExchange，允许开发人员匹配来自 HTTP 请求的任何内容，例如，Header 或参数。

过滤器（Filter）是使用特定工厂构建的 Spring Framework Gateway Filter 实例，可以在返回请求之前或之后修改请求和响应的内容。

Spring Cloud Gateway 集成 Hystrix 熔断器、集成 Spring Cloud DiscoveryClient、集成开发人员易于编写的 Predicates 和 Filters，提供了基于 Redis 的 Ratelimiter 实现，并使用令牌桶算法限流，支持路径重写，可以基于 Redis 使数据库实现动态路由。

7.2.2　Kotlin 集成 Spring Cloud Gateway

新建一个 Maven 子工程：chapter07-gateway，这是一个基于 Spring Cloud Gateway 搭建的网关服务，pom 文件如下：

```xml
1.  <?xml version="1.0" encoding="UTF-8"?>
2.  <project xmlns="http://maven.apache.org/POM/4.0.0"
3.           xmlns:xsi="http://www.w3.org/2001/XMLSchema-instance"
4.           xsi:schemaLocation="http://maven.apache.org/POM/4.0.0
    http://maven.apache.org/xsd/maven-4.0.0.xsd">
5.      <parent>
6.          <artifactId>kotlinspringboot</artifactId>
7.          <groupId>io.kang.kotlinspringboot</groupId>
8.          <version>0.0.1-SNAPSHOT</version>
9.      </parent>
10.     <modelVersion>4.0.0</modelVersion>
11.     <!-- 子工程名-->
12.     <artifactId>chapter07-gateway</artifactId>
13.     <dependencies>
14.         <!-- Spring Cloud Gateway 依赖包-->
15.         <dependency>
```

```xml
16.            <groupId>org.springframework.cloud</groupId>
17.            <artifactId>spring-cloud-starter-gateway</artifactId>
18.            <version>2.2.1.RELEASE</version>
19.        </dependency>
20.        <!-- Spring Cloud Eureka 客户端依赖包-->
21.        <dependency>
22.            <groupId>org.springframework.cloud</groupId>
23.            <artifactId>spring-cloud-starter-netflix-eureka-client</artifactId>
24.            <version>2.2.1.RELEASE</version>
25.        </dependency>
26.        <!-- Spring Boot Redis 依赖包-->
27.        <dependency>
28.            <groupId>org.springframework.boot</groupId>
29.            <artifactId>spring-boot-starter-data-redis-reactive</artifactId>
30.            <version>2.2.4.RELEASE</version>
31.        </dependency>
32.        <!-- Spring Cloud Hystrix 依赖包-->
33.        <dependency>
34.            <groupId>org.springframework.cloud</groupId>
35.            <artifactId>spring-cloud-starter-netflix-hystrix</artifactId>
36.            <version>2.2.1.RELEASE</version>
37.        </dependency>
38.        <!-- Spring Boot Webflux 依赖包-->
39.        <dependency>
40.            <groupId>org.springframework.boot</groupId>
41.            <artifactId>spring-boot-starter-webflux</artifactId>
42.            <version>2.2.1.RELEASE</version>
43.        </dependency>
44.        <!-- Fastjson 依赖包-->
45.        <dependency>
46.            <groupId>com.alibaba</groupId>
47.            <artifactId>fastjson</artifactId>
48.            <version>1.2.62</version>
49.        </dependency>
50.        <!-- Spring Boot Actuator 依赖包-->
51.        <dependency>
52.            <groupId>org.springframework.boot</groupId>
```

```xml
53.            <artifactId>spring-boot-starter-actuator</artifactId>
54.            <version>2.2.1.RELEASE</version>
55.        </dependency>
56.        <dependency>
57.            <groupId>com.fasterxml.jackson.module</groupId>
58.            <artifactId>jackson-module-kotlin</artifactId>
59.        </dependency>
60.        <dependency>
61.            <groupId>org.jetbrains.kotlin</groupId>
62.            <artifactId>kotlin-reflect</artifactId>
63.        </dependency>
64.        <dependency>
65.            <groupId>org.jetbrains.kotlin</groupId>
66.            <artifactId>kotlin-stdlib-jdk8</artifactId>
67.        </dependency>
68.        <dependency>
69.            <groupId>org.jetbrains.kotlinx</groupId>
70.            <artifactId>kotlinx-coroutines-core</artifactId>
71.            <version>1.3.2</version>
72.        </dependency>
73.        <!-- Spring Boot Test 依赖包-->
74.        <dependency>
75.            <groupId>org.springframework.boot</groupId>
76.            <artifactId>spring-boot-starter-test</artifactId>
77.            <version>2.2.1.RELEASE</version>
78.        </dependency>
79.    </dependencies>
80.    <build>
81.        <sourceDirectory>${project.basedir}/src/main/kotlin</sourceDirectory>
82.        <testSourceDirectory>${project.basedir}/src/test/kotlin</testSourceDirectory>
83.        <plugins>
84.            <plugin>
85.                <groupId>org.springframework.boot</groupId>
86.                <artifactId>spring-boot-maven-plugin</artifactId>
87.            </plugin>
88.            <plugin>
89.                <groupId>org.jetbrains.kotlin</groupId>
```

```xml
90.            <artifactId>kotlin-maven-plugin</artifactId>
91.            <configuration>
92.                <args>
93.                    <arg>-Xjsr305=strict</arg>
94.                </args>
95.                <compilerPlugins>
96.                    <plugin>spring</plugin>
97.                    <plugin>jpa</plugin>
98.                </compilerPlugins>
99.            </configuration>
100.           <dependencies>
101.               <dependency>
102.                   <groupId>org.jetbrains.kotlin</groupId>
103.                   <artifactId>kotlin-maven-allopen</artifactId>
104.                   <version>${kotlin.version}</version>
105.               </dependency>
106.               <dependency>
107.                   <groupId>org.jetbrains.kotlin</groupId>
108.                   <artifactId>kotlin-maven-noarg</artifactId>
109.                   <version>${kotlin.version}</version>
110.               </dependency>
111.           </dependencies>
112.        </plugin>
113.     </plugins>
114.  </build>
115. </project>
```

application.yml 文件中的定义如下：定义了对 provide-server 服务的路由映射，/provide/** 的路径会被转发到 provide-server 的 "/**" 接口；StripPrefix=1 表示忽略第一个路径，即 /provide。定义了一个服务不可用，熔断的默认接口为 "/fallback"。定义了一个限流器，其根据 apiKeyResolver 定义的 Bean 的名字，对相应的 bean 的调用进行限流：

```yaml
1. server:
2.   port: 8071    #应用端口号
3. spring:
4.   application:
5.     name: gateway-app    #应用名
6.   cloud:
7.     gateway:
8.       discovery:
```

```yaml
9.        locator:
10.          enabled: true # 是否和服务注册与发现组件结合，设置为 true 后可以直接使用应
                          # 用名称调用服务
11.      routes:
12.        - id: provide-server   #被路由代理的服务 id
13.          uri: lb://provider-server  #被路由代理的 uri
14.          predicates:
15.            - Path=/provide/**  #被代理的 url
16.          filters:
17.            ## 截取路径位数
18.            - StripPrefix=1
19.            - name: Hystrix   #Hystrix 配置
20.              args:
21.                name: fallBackBean  #熔断器名称
22.                fallbackUri: forward:/fallback  #提供熔断的接口
23.            - name: RequestRateLimiter  #限流器配置
24.              args:
25.                ### 限流过滤器的 Bean 名称
26.                key-resolver: '#{@apiKeyResolver}'
27.                ### 希望允许用户每秒处理多少个请求
28.                redis-rate-limiter.replenishRate: 1
29.                ### 用户允许在 1 秒内完成的最大请求数
30.                redis-rate-limiter.burstCapacity: 3
31.  redis:
32.    host: localhost  #Redis 连接 host
33.    prot: 6379    #Redis 连接端口
34.    password: 123456  #Redis 连接密码
35. eureka:
36.   client:
37.     service-url:
38.       defaultZone: http://localhost:8761/eureka/  #Eureka 注册中心地址
```

GateWayApplication.kt 定义了一个启动类，服务启动后会注册到 Eureka：

```kotlin
1. @SpringBootApplication
2. @EnableEurekaClient
3. class GateWayApplication
4. // 启动函数
5. fun main(args: Array<String>) {
```

```
6.      runApplication<GateWayApplication>(*args)
7.  }
```

顺序启动 chapter05-eureka、chapter05-eureka-provider、chapter07-gateway，调用 chapter07-gateway 的 "/provide/provide" 可以将请求转发到 chapter05-eureka-provider 的 "provide" 接口。

FallbackController.kt 定义了服务不可用时默认的接口：

```
1.  class FallbackController {
2.      // 提供熔断服务的接口
3.      @GetMapping("/fallback")
4.      fun fallback(): String {
5.          return "I'm Spring Cloud Gateway fallback."
6.      }
7.  }
```

关闭 chapter05-eureka-provider 服务，再次调用 "/provide/provide"，会转到 "/fallback" 接口，返回 "I'm Spring Cloud Gateway fallback."，如图7.6所示。

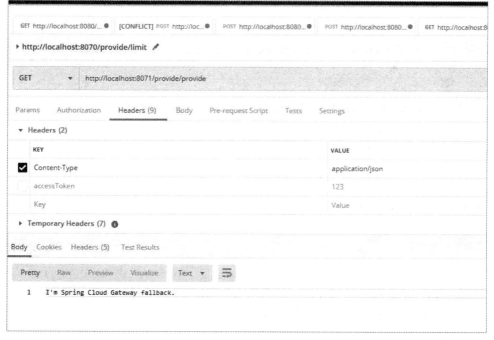

图7.6 /provide/provide接口服务降级调用的结果

RequestRateLimiterConfig.kt 定义了几种限流策略。ipAddressKeyResolver 根据调用服务器的主机名限流，apiKeyResolver 根据调用的路径进行限流，userKeyResolver 根据查询参数 userId 进行限流，yml 文件定义的限流策略是 apiKeyResolver：

```kotlin
1.  @Configuration
2.  class RequestRateLimiterConfig {
3.      // 根据主机名对请求进行限流
4.      @Bean("ipAddressKeyResolver")
5.      fun ipAddressKeyResolver(): KeyResolver {
6.          return KeyResolver {
7.              exchange -> Mono.just(exchange.request.remoteAddress?.hostName.orEmpty())
8.          }
9.      }
10.     // 根据调用路径对请求进行限流
11.     @Bean("apiKeyResolver")
12.     @Primary
13.     fun apiKeyResolver(): KeyResolver {
14.         return KeyResolver {
15.             exchange -> Mono.just(exchange.request.path.value())
16.         }
17.     }
18.     // 根据请求参数的 userId 对请求进行限流
19.     @Bean("userKeyResolver")
20.     fun userKeyResolver(): KeyResolver {
21.         return KeyResolver {
22.             exchange -> Mono.just(exchange.request.queryParams.getFirst("userId").orEmpty())
23.         }
24.     }
25. }
```

使用 postman 对接口 "/provide/provide" 做并发测试，并发数为 10，只有 4 次调用返回正常，其他调用都被限制了，如图 7.7 所示。

图7.7　provide/provide接口并发限流测试结果

RedisRouteRepository.kt 将路由缓存到 Redis，实现对路由配置的动态刷新。save 方法可以新增路由配置，getRouteDefinitions 方法返回所有的路由配置，delete 方法根据路由 id 删除路由配置。代码如下：

```kotlin
@Component
class RedisRouteRepository: RouteDefinitionRepository {
    // 注入 redisTemplate
    @Autowired
    lateinit var redisTemplate: RedisTemplate<String, String>
    val routeKey = "route"
    // 保存路由配置
    override fun save(route: Mono<RouteDefinition>): Mono<Void> {
        route
            .subscribe { routeDefinition ->
                println("${JSON.toJSONString(routeDefinition)}")
                redisTemplate.opsForHash<String, String>().put(
                    routeKey, routeDefinition.id, JSON.toJSONString(routeDefinition))
            }
        return Mono.empty<Void>()
    }
    // 获取缓存中所有的路由配置
```

```kotlin
18.     override fun getRouteDefinitions(): Flux<RouteDefinition> {
19.         if (redisTemplate.hasKey(routeKey)) {
20.             // 从 Redis 中拉取路由
21.             val routeDefinitions = LinkedList<RouteDefinition>()
22.             redisTemplate
23.                     .opsForHash<String, String>().values(routeKey)
24.                     .stream()
25.                     .forEach { routeDefinition ->
    routeDefinitions.add(JSON.parseObject(routeDefinition,
    RouteDefinition::class.java) as RouteDefinition) }
26.             return Flux.fromIterable(routeDefinitions)
27.         } else {
28.             var routes = LinkedHashMap<String, String>()
29.             redisTemplate.opsForHash<String, String>().putAll(routeKey, routes)
30.             return Flux.fromIterable(LinkedList<RouteDefinition>())
31.         }
32.     }
33.     // 根据 routeId 从缓存中删除路由配置
34.     override fun delete(routeId: Mono<String>): Mono<Void> {
35.         routeId
36.                 .subscribe { routeId ->
37.                     if (redisTemplate.opsForHash<String,
    String>().hasKey(routeKey, routeId)) {
38.                         redisTemplate.opsForHash<String,
    String>().delete(routeKey, routeId)
39.                     }
40.                 }
41.         return Mono.empty<Void>()
42.     }
43. }
```

RouteController.kt 定义了添加、查询、删除路由的接口：

```kotlin
1. @RestController
2. @RequestMapping("/route")
3. class RouteController {
4.     @Autowired
5.     lateinit var redisRouteRepository: RedisRouteRepository
```

```kotlin
6.    // 测试接口，添加路由
7.    @PostMapping("/add")
8.    fun add(@RequestBody routeDefinition: RouteDefinition): Mono<String> {
9.        redisRouteRepository.save(Mono.just(routeDefinition))
10.       return Mono.just("add ok")
11.   }
12.   // 测试接口，获取所有路由
13.   @GetMapping("/all")
14.   fun getAll(): Flux<RouteDefinition> {
15.       return redisRouteRepository.routeDefinitions
16.   }
17.   // 测试接口，删除指定 id 路由
18.   @DeleteMapping("/{id}")
19.   fun delete(@PathVariable id: String): Mono<String> {
20.       redisRouteRepository.delete(Mono.just(id))
21.       return Mono.just("delete ok")
22.   }
23. }
```

启动 chapter05-eureka-consumer 工程，使用 postman 调用 "/route/add" 接口添加一个新的路由配置，如图 7.8 所示。

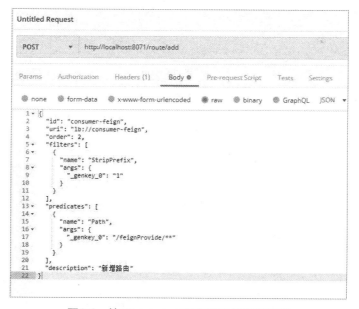

图7.8　接口route/add添加路由调用示意图

增加了对 consumer-feign 这个服务的路由配置,调用"/feignProvide/feignProvide"接口,可以将请求转发到 consumer-feign 的"feignProvide"接口。调用"/route/consumer-feign"接口,可删除该条配置,再次调用"/feignProvide/feignProvide",调用不通。通过 Redis,可以动态更新缓存配置,实现路由配置热加载。

7.3 小结

网关屏蔽内部细节,为调用者提供统一入口。网关接收所有调用者请求,并通过路由机制转发到服务实例。网关是一组"过滤器"集合,可以实现一系列与核心业务无关的横切面功能,如安全认证、限流熔断、日志监控等。

本节介绍了两种网关:Zuul 和 Spring Cloud Gateway。Spring Cloud 对它们都提供了很好的集成,使用起来很方便。Zuul1 基于 Servlet,Zuul2 基于 Netty,是非阻塞的、响应式的;Spring Cloud Gateway 基于 Reactor,是异步非阻塞的。Zuul1 采用同步阻塞方式,性能较差,Spring Cloud Gateway 和 Zuul2 性能较好。

第 8 章
Kotlin 应用于 Spring Cloud Alibaba

Spring Cloud Alibaba 是阿里巴巴集团开源的微服务开发的一站式解决方案。该项目包含开发分布式应用服务的必备组件，方便开发者通过 Spring Cloud 编程模型轻松使用这些组件来开发分布式应用服务。依托 Spring Cloud Alibaba，只需添加一些注解和少量配置，就可以将 Spring Cloud 应用接入阿里巴巴分布式应用解决方案，通过阿里巴巴中间件来迅速搭建分布式应用系统。

Spring Cloud Alibaba 包含开源组件和商业化组件。开源组件包括：分布式配置管理、服务注册与发现、服务限流降级、消息中间件、Dubbo Spring Cloud 及分布式事务。商业化组件包括：商业版服务注册与发现、商业版应用配置管理、对象存储服务、分布式任务调度及通信服务。

下面详细介绍其中的一些组件。

服务限流降级：默认支持 WebServlet、WebFlux、OpenFeign、RestTemplate、Spring Cloud Gateway、Zuul、Dubbo 和 RocketMQ 等组件限流降级功能的接入，可以在运行时通过控制台实时修改限流降级规则，还支持查看限流降级指标监控。

服务注册与发现：适配 Spring Cloud 服务注册与发现标准，默认集成了 Ribbon 的支持。

分布式配置管理：支持分布式系统中的外部化配置，配置更改时自动刷新。

消息驱动能力：基于 Spring Cloud Stream 为微服务应用构建消息驱动能力。

分布式事务：使用@GlobalTransactional 注解，高效且对业务零侵入地解决分布式事务问题。

分布式任务调度：提供秒级、精准、高可靠、高可用的定时（基于 Cron 表达式）任务调度服务。同时提供分布式的任务执行模型，如网格任务。网格任务支持将海量子任务均匀分配到所有 Worker（schedulerx-client）上执行。

阿里云短信服务：覆盖全球的短信服务，友好、高效、智能的互联化通信能力，帮助企业迅速搭建客户触达通道。

Sentinel：把流量管理作为切入点，从流量控制、熔断降级、系统负载保护等多个维度保护服务的稳定性。

Nacos：提供了一个易于构建云原生应用的动态服务发现、配置管理和服务管理平台。

RocketMQ：一款开源的分布式消息系统，基于高可用分布式集群技术，提供低延时的、高可靠的消息发布与订阅服务。

Dubbo：一款高性能的 Java RPC 框架。

Seata：一个易于使用的高性能微服务分布式事务解决方案。

Alibaba Cloud ACM：一款在分布式架构环境中对应用配置进行集中管理和推送的应用配置中心产品。

Alibaba Cloud OSS：阿里云对象存储服务（Object Storage Service，简称 OSS），提供海量、安全、低成本、高可靠的云存储服务。可以在任何应用、任何时间、任何地点存储和访问任意类型的数据。

8.1 服务限流降级

Sentinel 诞生于 2012 年，其主要功能是控制接口流量。2013—2017 年，Sentinel 在阿里巴巴集团内部使用，2018 年，Sentinel 被开源。本节主要介绍 Kotlin 集成 Sentinel 实现服务限流、降级的方法。

8.1.1　Sentinel 介绍

资源是 Sentinel 的关键概念，它可以是 Java 应用程序中的任何内容，例如，由应用程序提供的服务，或由应用程序调用其他应用提供的服务，甚至是一段代码。

规则是围绕资源的实时状态设定的，包括流量控制规则、熔断降级规则、系统负载保护规则，所有规则都可以动态实时调整。

流量控制用于调整网络包发送的数据，包括：资源的调用关系，例如，资源的调用链路、资源和资源之间的关系；运行指标，例如，QPS、线程池、系统负载等；控制的方式，例如，直接限流、冷启动、排队。Sentinel 作为一个调配器，可以根据需要把随机的请求调整成合适的形状，如图 8.1 所示。

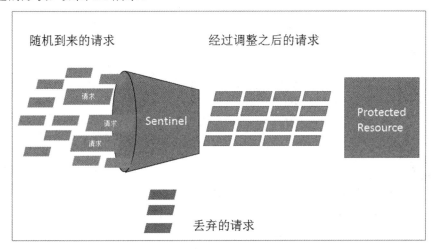

图8.1　Sentinel功能示意图

Sentinel 和 Hystrix 的原则一致，当调用链路中某个资源出现不稳定情况时，例如超时、异常比例升高，对这个资源的调用进行限制，并让请求快速失败，避免影响到其他资源。在限制的手段上，Sentinel 和 Hystrix 采取了完全不同的方法。

Hystrix 通过线程池的方式对依赖（对应 Sentinel 中的资源）进行隔离。这样做的好处是资源和资源之间做到了彻底隔离；缺点是增加了线程切换的成本，需要预先给各个资源进行线程池大小的分配。

Sentinel 通过并发线程数进行限制，限制资源并发线程的数量来减少不稳定资源对其他资源的影响。这样不但可减少线程切换的损耗，也不需要预先分配线程池大小。当某个资源出现不稳定的情况时，例如，响应时间变长，会造成线程数的逐步堆积。当线程数在特定资源上堆积到一定数量之后，对该资源的新请求就会被拒绝。堆积的线程完成任务后才会开始继续接收请求。此外，Sentinel 通过响应时间对资源进行降级，通过响应时间来快速降级不稳定的资源。当依赖的资源出现响应时间过长时，所有对资源的访问都会被直接拒绝，直到过了指定的时间之后才会重新恢复。

Sentinel 同时对系统的维度提供保护，防止"雪崩"。当系统负载较高时，如果还持续让请求进入，可能会导致系统崩溃、无法响应。在集群环境下，网络负载均衡会把本应这台机器承载的流量转发到其他机器上。如果这时那个"其他机器"也处于边缘状态，那么增加的流量就会导致这台机器也崩溃，最后导致整个集群不可用。

8.1.2 Kotlin 集成 Sentinel

新建一个 Maven 子工程：chapter08-sentinel，使用 Sentinel 对接口进行限流。pom 文件如下：

```
1.  <?xml version="1.0" encoding="UTF-8"?>
2.  <project xmlns="http://maven.apache.org/POM/4.0.0"
3.           xmlns:xsi="http://www.w3.org/2001/XMLSchema-instance"
4.           xsi:schemaLocation="http://maven.apache.org/POM/4.0.0
    http://maven.apache.org/xsd/maven-4.0.0.xsd">
5.      <parent>
6.          <artifactId>kotlinspringboot</artifactId>
7.          <groupId>io.kang.kotlinspringboot</groupId>
8.          <version>0.0.1-SNAPSHOT</version>
9.      </parent>
10.     <modelVersion>4.0.0</modelVersion>
11.     <!-- 子工程名 -->
12.     <artifactId>chapter08-sentinel</artifactId>
13.     <dependencies>
14.         <!-- Spring Cloud Sentinel 依赖包-->
15.         <dependency>
16.             <groupId>com.alibaba.cloud</groupId>
17.             <artifactId>spring-cloud-starter-alibaba-sentinel</artifactId>
18.             <version>2.2.0.RELEASE</version>
19.         </dependency>
20.         <!-- Spring Boot Web 依赖包-->
21.         <dependency>
```

```xml
22.            <groupId>org.springframework.boot</groupId>
23.            <artifactId>spring-boot-starter-web</artifactId>
24.            <version>2.2.1.RELEASE</version>
25.        </dependency>
26.        <!-- Sentinel Nacos 依赖包-->
27.        <dependency>
28.            <groupId>com.alibaba.csp</groupId>
29.            <artifactId>sentinel-datasource-nacos</artifactId>
30.            <version>1.7.1</version>
31.        </dependency>
32.        <dependency>
33.            <groupId>com.fasterxml.jackson.module</groupId>
34.            <artifactId>jackson-module-kotlin</artifactId>
35.        </dependency>
36.        <dependency>
37.            <groupId>org.jetbrains.kotlin</groupId>
38.            <artifactId>kotlin-reflect</artifactId>
39.        </dependency>
40.        <dependency>
41.            <groupId>org.jetbrains.kotlin</groupId>
42.            <artifactId>kotlin-stdlib-jdk8</artifactId>
43.        </dependency>
44.        <dependency>
45.            <groupId>org.jetbrains.kotlinx</groupId>
46.            <artifactId>kotlinx-coroutines-core</artifactId>
47.            <version>1.3.2</version>
48.        </dependency>
49.    </dependencies>
50.    <build>
51.        <sourceDirectory>${project.basedir}/src/main/kotlin</sourceDirectory>
52.        <testSourceDirectory>${project.basedir}/src/test/kotlin</testSourceDirectory>
53.        <plugins>
54.            <plugin>
55.                <groupId>org.springframework.boot</groupId>
56.                <artifactId>spring-boot-maven-plugin</artifactId>
57.            </plugin>
58.            <plugin>
```

```xml
59.         <groupId>org.jetbrains.kotlin</groupId>
60.         <artifactId>kotlin-maven-plugin</artifactId>
61.         <configuration>
62.             <args>
63.                 <arg>-Xjsr305=strict</arg>
64.             </args>
65.             <compilerPlugins>
66.                 <plugin>spring</plugin>
67.                 <plugin>jpa</plugin>
68.             </compilerPlugins>
69.         </configuration>
70.         <dependencies>
71.             <dependency>
72.                 <groupId>org.jetbrains.kotlin</groupId>
73.                 <artifactId>kotlin-maven-allopen</artifactId>
74.                 <version>${kotlin.version}</version>
75.             </dependency>
76.             <dependency>
77.                 <groupId>org.jetbrains.kotlin</groupId>
78.                 <artifactId>kotlin-maven-noarg</artifactId>
79.                 <version>${kotlin.version}</version>
80.             </dependency>
81.         </dependencies>
82.     </plugin>
83.   </plugins>
84. </build>
85. </project>
```

application.yml 文件的内容如下：

```yml
1. server:
2.   port: 8081 #应用端口号
3. spring:
4.   application:
5.     name: sentinel-app #应用名
6.   cloud:
7.     sentinel:
```

```
8.      datasource:
9.        ds1:
10.         file:
11.           file: classpath:flowrule.json  #限流配置文件名
12.           data-type: json  #文件格式
13.           rule-type: flow  #限流规则类型：以 JSON 格式返回现有的限流规则
14.        ds2:
15.          nacos:
16.           server-addr: localhost:8848  #Nacos 配置中心地址
17.           data-id: ${spring.application.name}-rule  #Nacos 配置的 data-id
18.           group-id: DEFAULT_GROUP  #Nacos 配置的 group
19.           rule-type: flow  #限流规则类型：以 JSON 格式返回现有的限流规则
20.     transport:
21.       dashboard: localhost:8080  #应用端口号
22.       eager: true  #应用端口号
```

Sentinel 的各种保护规则可以存储在 JSON 文件中，也可以存储在 Nacos 等配置中心。application.yml 定义了两种存放规则的方式：flowrule.json 和 nacos。flowrule.json 文件如下：

```
1.  [
2.    {
3.      "resource": "helloApi",
4.      "controlBehavior": 1,
5.      "count": 10,
6.      "grade": 1,
7.      "limitApp": "default",
8.      "strategy": 0
9.    },
10.   {
11.     "resource": "helloApi1",
12.     "controlBehavior": 2,
13.     "count": 10,
14.     "grade": 1,
15.     "limitApp": "default",
16.     "strategy": 0
17.   }
18. ]
```

resource 代表资源,是限流规则的作用对象;count 是限流阈值;grade 是限流阈值类型,0 表示基于线程数,1 表示基于 QPS;limitApp 是流量控制针对的调用来源,default 表示不区分调用来源;controlBehavior 是流量控制方式,0 表示默认,1 表示冷启动,2 表示匀速排队等待,3 表示慢启动;strategy 表示根据什么限流,0 表示根据资源本身,1 表示关联其他资源,2 表示根据链路入口。

Nacos 配置的限流规则如图 8.2 所示。

图8.2　Nacos配置的限流规则

SentinelApplication.kt 定义了一个启动类,启动一个 Spring Boot 应用:

```
1.  @SpringBootApplication
2.  class SentinelApplication {
3.      @Bean
4.      fun sentinelResourceAspect(): SentinelResourceAspect {
5.          return SentinelResourceAspect()
6.      }
7.  }
8.  // 启动函数
```

```kotlin
9.  fun main(args: Array<String>) {
10.     runApplication<SentinelApplication>(*args)
11. }
```

SentinelController.kt 定义了三个接口，用于测试限流规则：

```kotlin
1.  @RestController
2.  class SentinelController {
3.      // 测试接口，被限流后默认调用 handleException 函数
4.      @GetMapping("/hello")
5.      @SentinelResource(value = "helloApi", blockHandler = "handleException",
    blockHandlerClass = [ExceptionUtil::class])
6.      fun hello(): String {
7.          return "hello sentinel"
8.      }
9.      // 测试接口，被限流后默认调用 exceptionHandler 函数
10.     @GetMapping("/hello1")
11.     @SentinelResource(value = "helloApi1", blockHandler = "exceptionHandler")
12.     fun hello1(): String {
13.         return "hello1 sentinel"
14.     }
15.     // 测试接口，从 Nacos 读取限流规则，被限流后默认调用 exceptionHandler 函数
16.     @GetMapping("/hello2")
17.     @SentinelResource(value = "helloNacosApi", blockHandler = "exceptionHandler")
18.     fun helloNacos(): String {
19.         return "hello1 sentinel in nacos"
20.     }
21.     //被限流后默认调用 exceptionHandler 函数
22.     fun exceptionHandler(s: Long, ex: BlockException): String {
23.         println(ex.printStackTrace())
24.         return "Oops, error occurred at $s"
25.     }
26. }
```

ExceptionUtil.kt 定义异常处理方法：

```kotlin
1.  // 异常处理方法
2.  class ExceptionUtil {
3.      fun handleException(ex: BlockException): String {
```

```
4.        return "Oops:${ex.javaClass.canonicalName}"
5.    }
6. }
```

"/hello"和"/hello1"两个接口的限流规则存放在 flowrule.json 文件中，"/hello2"限流规则存放在 Nacos 中。将 Nacos 的 controlBehavior 改为 0，使用 postman 调用"/hello2"接口 10 次，每次调用都正常返回；将 controlBehavior 改为 1，调用"/hello2"接口 10 次，只有三次正常返回，其余 7 次调用被限制；将 controlBehavior 改为 2，调用"/hello2"接口 10 次，接口匀速排队，调用正常；将 controlBehavior 改为 3，调用"/hello2"接口 10 次，接口缓慢排队，调用正常，如图 8.3 所示。

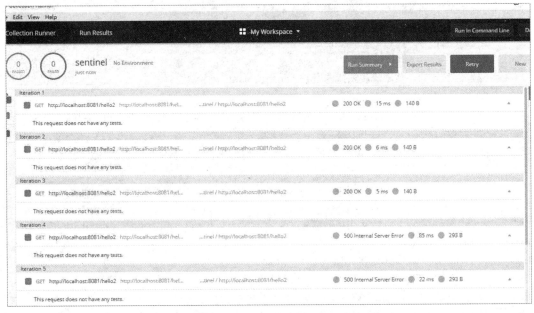

图8.3　接口hello2并发限流调用结果

8.2　消息驱动

消息驱动对于微服务系统很重要，有助于系统间解耦。Spring Cloud Stream 是一个为微服务应用构建消息驱动能力的框架。通过使用 Spring Cloud Stream，可以有效降低消息中间件的使用复杂度，让系统开发人员可以有更多的精力关注于核心业务逻辑的处理。本节介绍使用 Kotlin 集成 Spring Cloud Stream、RocketMQ 开发的相关知识。

8.2.1 消息驱动介绍

Spring Cloud Stream 为一些消息中间件提供了个性化的自动化配置功能，并引入了发布-订阅、消费组、分区这三个核心概念。RocketMQ 可以和 Spring Cloud Stream 集成。

应用程序通过 inputs、outputs 与 Spring Cloud Stream 中的 Binder 交互，Spring Cloud Stream 中的 Binder 负责与消息中间件交互。只需要搞清楚如何与 Spring Cloud Stream 交互，就可以方便使用消息驱动的方式，如图 8.4 所示。

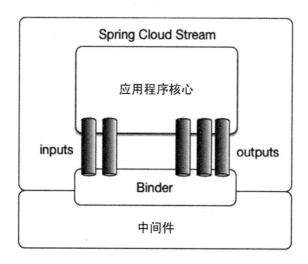

图8.4 Spring Cloud Stream交互示意图

通过使用 Spring Integration 连接消息代理中间件就可以实现消息事件驱动。企业应用集成是一种应用之间数据和服务集成的技术，有如下集成风格。

- 共享数据库，两个系统查询同一个数据库以获取要传递的数据。例如，部署了两个 Spring Boot 应用，它们的实体类共用同一个表。
- 远程过程调用，两个系统都暴露另一个能调用的服务。例如，EJB 服务、SOAP 或 REST 服务。
- 消息，两个系统连接到一个公用的消息系统，互相交换数据，并利用消息调用行为。例如，中心辐射式的（hub-and-spoke）JMS 架构。

Spring Cloud Stream 提供了一种应用和消息中间件解耦的方式。Spring Cloud Stream 由一个中间件中立的核组成。inputs，相当于消费者 consumer，它从队列中接收消息；outputs，相当于生产者 producer，它从队列中发送消息。应用通过 inputs 和 outputs 通道与外界交流。

通道通过指定中间件的 Binder 实现与外部代理连接。业务开发者不用关注具体的消息中间件，只需关注 Binder 对应用程序提供的抽象概念以使用消息中间件实现业务即可。

Binder 通过定义绑定器，并将其作为中间层，实现了应用程序与消息中间件细节之间的隔离。通过向应用程序暴露统一的 Channel 通道，使应用程序不需要考虑各种不同的消息中间件的实现细节。当需要升级消息中间件，或者更换其他消息中间件产品时，更换对应的 Binder 绑定器即可，不需要修改任何应用逻辑，甚至可以任意改变中间件的类型而不需要修改一行代码。阿里巴巴提供了 RocketMQ 的 Binder 实现——spring-cloud-stream-binder-rocketmq。

8.2.2 Kotlin 集成 RocketMQ 实现消息驱动

新建 Maven 子工程：chapter08-stream-rocketmq，这是一个基于 RocketMQ 的 Spring Cloud Stream 应用，pom 文件如下：

```xml
<?xml version="1.0" encoding="UTF-8"?>
<project xmlns="http://maven.apache.org/POM/4.0.0"
    xmlns:xsi="http://www.w3.org/2001/XMLSchema-instance"
    xsi:schemaLocation="http://maven.apache.org/POM/4.0.0 http://maven.apache.org/xsd/maven-4.0.0.xsd">
    <parent>
        <artifactId>kotlinspringboot</artifactId>
        <groupId>io.kang.kotlinspringboot</groupId>
        <version>0.0.1-SNAPSHOT</version>
    </parent>
    <modelVersion>4.0.0</modelVersion>
    <!-- 子工程名-->
    <artifactId>chapter08-stream-rocketmq</artifactId>
    <dependencies>
        <!-- Spring Boot Web 依赖包-->
        <dependency>
            <groupId>org.springframework.boot</groupId>
            <artifactId>spring-boot-starter-web</artifactId>
            <version>2.2.1.RELEASE</version>
        </dependency>
        <!-- Spring Cloud RocketMQ 依赖包-->
        <dependency>
```

```xml
22.            <groupId>com.alibaba.cloud</groupId>
23.            <artifactId>spring-cloud-stream-binder-rocketmq</artifactId>
24.            <version>2.2.0.RELEASE</version>
25.        </dependency>
26.        <dependency>
27.            <groupId>com.fasterxml.jackson.module</groupId>
28.            <artifactId>jackson-module-kotlin</artifactId>
29.        </dependency>
30.        <dependency>
31.            <groupId>org.jetbrains.kotlin</groupId>
32.            <artifactId>kotlin-reflect</artifactId>
33.        </dependency>
34.        <dependency>
35.            <groupId>org.jetbrains.kotlin</groupId>
36.            <artifactId>kotlin-stdlib-jdk8</artifactId>
37.        </dependency>
38.        <dependency>
39.            <groupId>org.jetbrains.kotlinx</groupId>
40.            <artifactId>kotlinx-coroutines-core</artifactId>
41.            <version>1.3.2</version>
42.        </dependency>
43.    </dependencies>
44.    <build>
45.        <sourceDirectory>${project.basedir}/src/main/kotlin</sourceDirectory>
46.        <testSourceDirectory>${project.basedir}/src/test/kotlin</testSourceDirectory>
47.        <plugins>
48.            <plugin>
49.                <groupId>org.springframework.boot</groupId>
50.                <artifactId>spring-boot-maven-plugin</artifactId>
51.            </plugin>
52.            <plugin>
53.                <groupId>org.jetbrains.kotlin</groupId>
54.                <artifactId>kotlin-maven-plugin</artifactId>
55.                <configuration>
56.                    <args>
57.                        <arg>-Xjsr305=strict</arg>
58.                    </args>
```

```xml
59.            <compilerPlugins>
60.                <plugin>spring</plugin>
61.                <plugin>jpa</plugin>
62.            </compilerPlugins>
63.         </configuration>
64.         <dependencies>
65.            <dependency>
66.                <groupId>org.jetbrains.kotlin</groupId>
67.                <artifactId>kotlin-maven-allopen</artifactId>
68.                <version>${kotlin.version}</version>
69.            </dependency>
70.            <dependency>
71.                <groupId>org.jetbrains.kotlin</groupId>
72.                <artifactId>kotlin-maven-noarg</artifactId>
73.                <version>${kotlin.version}</version>
74.            </dependency>
75.         </dependencies>
76.        </plugin>
77.      </plugins>
78.   </build>
79. </project>
```

application.yml 文件的内容如下：

```
1.  server:
2.    port: 8082    #应用端口
3.  spring:
4.    application:
5.      name: rocketmq-stream    #应用名
6.    cloud:
7.      stream:
8.        rocketmq:
9.          binder:
10.           name-server: localhost:9876    #RocketMQ name-server 地址
11.        bindings:
12.          input1:
13.            consumer:
```

```yaml
14.          orderly: true    #顺序消费
15.          tags: tagStr0    #消息 tag
16.        input2:
17.          consumer:
18.            orderly: false   #不保证消费顺序
19.            tags: tagStr1    #消息 tag
20.      bindings:
21.        output1:
22.          destination: test-topic    #topic 名称
23.          content-type: text/plain   #消息格式
24.          producer:
25.            partitionKeyExpression: headers['partitionKey']   #分区
26.            partitionCount: 2   #分区个数
27.        input1:
28.          destination: test-topic    #topic 名称
29.          content-type: text/plain   #消息格式
30.          group: test-consumer-group1   #消费组
31.          consumer:
32.            instance-index: 0   #用哪个分区来接收消息
33.            instance-count: 2   #分区数
34.        input2:
35.          destination: test-topic    #topic 名称
36.          content-type: text/plain   #消息格式
37.          group: test-consumer-group2   #消费组
38.          consumer:
39.            concurrency: 20   #并行度
40.            instance-index: 1   #用哪个分区来接收消息
41.            instance-count: 2   #分区数
```

application.yml 定义了一个 output 和两个 input。output 相当于生产者，input 相当于消费者。output1 向两个分区发送消息，分区根据"partitionKey"进行区分。input1 属于消费组"test-consumer-group1"，接收分区 0 的消息，接收 tag 是"tagStr0"的消息。input2 属于消费组"test-consumer-group2"，接收分区 1 的消息，接收 tag 是"tagStr1"的消息。这两个消费组都可以消费"test-topic"的消息，相当于广播模式。同一个消费组有多个实例，共同使用（消费）topic 的消息，每条消息只会被一个消费组使用。

StreamApplication.kt 定义了一个启动类，注入自定义的输入、输出类：

```kotlin
// 开启 Spring Cloud Stream Binding 注解
@SpringBootApplication
@EnableBinding(MySource::class, MySink::class)
class StreamApplication
// 启动函数
fun main(args: Array<String>) {
    runApplication<StreamApplication>(*args)
}
```

MySink.kt 定义了两个输入,即两个消费者:

```kotlin
interface MySink {
    // 输入 1
    @Input("input1")
    fun input1(): SubscribableChannel
    // 输入 2
    @Input("input2")
    fun input2(): SubscribableChannel
}
```

MySource.kt 定义了一个输出,即一个生产者:

```kotlin
interface MySource {
    // 输出 1
    @Output("output1")
    fun output1(): MessageChannel
}
```

ReceiveService.kt 定义了两个消费者的行为,input1 和 input2 收到消息后分别打印消息内容:

```kotlin
@Service
class ReceiveService {
    // 监听 input1,收到消息,进行打印
    @StreamListener("input1")
    fun receiveInput1(receiveMsg: String) {
        println("input1 receive: $receiveMsg")
    }
    // 监听 input2,收到消息,进行打印
```

```
9.      @StreamListener("input2")
10.     fun receiveInput2(receiveMsg: String) {
11.         println("input2 receive: $receiveMsg")
12.     }
13. }
```

SteamController.kt 定义了三个接口，用于测试。"/send0"接口向第 0 个分区发送消息，消息的 tag 是"tagStr0"。"/send1"接口向第 1 个分区发送消息，消息的 tag 是"tagStr1"。"/send2"接口向第 1 个分区发送消息，消息的 tag 是"tagStr2"。代码如下：

```
1.  @RestController
2.  class SteamController {
3.      @Autowired
4.      lateinit var mySource: MySource
5.      // 测试接口，向第 0 个分区发送 tag 是 tagStr0 的消息
6.      @GetMapping("/send0/{id}")
7.      fun send(@PathVariable id: String) {
8.          val headers = HashMap<String, Any>()
9.          headers[MessageConst.PROPERTY_TAGS] = "tagStr0"
10.         headers["partitionKey"] = 0
11.         mySource.output1().send(MessageBuilder.createMessage("hello world: $id",
    MessageHeaders(headers)))
12.     }
13.     // 测试接口，向第 1 个分区发送 tag 是 tagStr1 的消息
14.     @GetMapping("/send1/{id}")
15.     fun send1(@PathVariable id: String) {
16.         val headers = HashMap<String, Any>()
17.         headers[MessageConst.PROPERTY_TAGS] = "tagStr1"
18.         headers["partitionKey"] = 1
19.         mySource.output1().send(MessageBuilder.createMessage("hello world: $id",
    MessageHeaders(headers)))
20.     }
21.     // 测试接口，向第 1 个分区发送 tag 是 tagStr2 的消息
22.     @GetMapping("/send2/{id}")
23.     fun send2(@PathVariable id: String) {
24.         val headers = HashMap<String, Any>()
25.         headers[MessageConst.PROPERTY_TAGS] = "tagStr2"
26.         headers["partitionKey"] = 1
```

```
27.        mySource.output1().send(MessageBuilder.createMessage("hello world: $id",
   MessageHeaders(headers)))
28.    }
29. }
```

调用"/send0/1",在控制台打印"input1 receive: hello world: 1",input1 成功消费这条消息。调用"/send1/1",在控制台打印"input2 receive: hello world: 1",input2 成功消费这条消息。调用"/send2/1",在控制台没有打印,因为 input2 只消费 tag 是"tagStr1"的消息,这条消息的 tag 是"tagStr2",被过滤掉了。通过图 8.5 可以看到,这条消息被"test-consumer-group1"和"test-consumer-group2"都过滤掉了。被"test-consumer-group1"过滤掉的原因是这条消息是第 1 个分区中的,而"test-consumer-group1"只消费发送到第 0 个分区的消息。被"test-consumer-group2"过滤掉的原因是"test-consumer-group2"只消费 tag 是"tagStr1"的消息。

图8.5 tag=tagStr2的消息的消费结果

8.3 阿里对象云存储

阿里对象云存储是阿里巴巴提供的对象存储 OSS（Object Storage Service）产品，方便用户存储图片资源，并通过 URL 访问图片。本节主要介绍 Kotlin 集成阿里云 OSS 的 SDK 进行文件上传及下载开发的方法。

8.3.1 阿里对象云存储介绍

为了更好地理解 OSS，这里介绍如下几个概念。

存储空间（Bucket）是用户存储对象（Object）的容器，所有的对象都必须隶属于某个存储空间。存储空间具有各种配置属性，包括地域、访问权限、存储类型等。用户可以根据实际需求，创建不同类型的存储空间来存储不同的数据。同一个存储空间的内部是扁平的，没有文件系统中的目录等概念，所有的对象都直接隶属于其对应的存储空间。每个用户可以拥有多个存储空间。存储空间的名称在 OSS 范围内必须是全局唯一的，一旦创建便无法修改名称。存储空间内部的对象数目没有限制。

对象/文件（Object）是 OSS 存储数据的基本单元，也被称为 OSS 的文件。对象由元信息（Object Meta）、用户数据（Data）和文件名（Key）组成。对象由存储空间内部唯一的 Key 来标识。对象元信息是一组键值对，表示对象的一些属性，比如最后修改时间、对象大小等信息，同时用户也可以在元信息中存储一些自定义的信息。对象的生命周期是从上传成功到被删除为止。在整个生命周期内，只有通过追加上传的对象可以继续通过追加上传写入数据，以其他上传方式上传的对象内容无法编辑，可以通过重复上传同名的对象来覆盖之前的对象。

地域（Region）表示 OSS 的数据中心所在的物理位置。用户可以根据费用、请求来源等选择合适的地域创建存储空间。一般来说，距离用户近的地域的访问速度更快。地域是在创建存储空间的时候指定的，一旦指定存储空间就不允许更改。该存储空间中所有的对象都存储在对应的数据中心，目前不支持对象级别的地域设置。

访问域名（Endpoint）表示 OSS 对外服务的访问域名。OSS 以 HTTP RESTful API 的形式对外提供服务，当访问不同地域的时候，需要不同的域名。通过内网和外网访问同一个地域所需要的访问域名也是不同的。

访问密钥（AccessKey）是访问身份验证中用到的 AccessKeyId 和 AccessKeySecret。OSS 通过使用 AccessKeyId 和 AccessKeySecret 对称加密的方法来验证某个请求的发送者身份。AccessKeyId 用于标识用户；AccessKeySecret 是用户加密签名字符串和 OSS 用来验证签名字符串的密钥,必须保密。

对象操作在 OSS 上具有原子性,OSS 保证用户一旦上传完成,读到的对象就是完整的,OSS 不会给用户返回一个"部分上传成功"的对象。对象操作在 OSS 上同样具有强一致性,用户一旦收到一个上传成功的响应,该上传的对象就已经立即可读,并且对象的冗余数据也同时写成功。

OSS 采用数据冗余存储机制,将每个对象的不同冗余存储在同一个区域内多个设施的多个设备上,以确保硬件失效时的数据可靠性和可用性。OSS 的冗余存储机制支持两个存储设施并发损坏时,仍可维持数据不丢失。当数据存入 OSS 后,OSS 会检测和修复丢失的冗余,确保数据可靠性和可用性。OSS 会周期性地通过校验等方式验证数据的完整性,及时发现因硬件失效等原因造成的数据损坏。当检测到数据有部分损坏或丢失时,OSS 会利用冗余的数据进行重建并修复损坏的数据。

8.3.2　Kotlin 集成阿里对象云存储

新建一个 Maven 子工程:chapter08-oss,通过 ossClient 访问阿里云对象存储。pom.xml 文件如下:

```
1.  <?xml version="1.0" encoding="UTF-8"?>
2.  <project xmlns="http://maven.apache.org/POM/4.0.0"
3.           xmlns:xsi="http://www.w3.org/2001/XMLSchema-instance"
4.           xsi:schemaLocation="http://maven.apache.org/POM/4.0.0
    http://maven.apache.org/xsd/maven-4.0.0.xsd">
5.      <parent>
6.          <artifactId>kotlinspringboot</artifactId>
7.          <groupId>io.kang.kotlinspringboot</groupId>
8.          <version>0.0.1-SNAPSHOT</version>
9.      </parent>
10.     <modelVersion>4.0.0</modelVersion>
11.     <!-- 子工程名 -->
12.     <artifactId>chapter08-oss</artifactId>
13.     <dependencies>
14.         <!-- Spring Boot Web 依赖包 -->
15.         <dependency>
16.             <groupId>org.springframework.boot</groupId>
17.             <artifactId>spring-boot-starter-web</artifactId>
```

```xml
18.            <version>2.2.1.RELEASE</version>
19.        </dependency>
20.        <!-- Spring Boot Cloud OSS 依赖包-->
21.        <dependency>
22.            <groupId>com.alibaba.cloud</groupId>
23.            <artifactId>spring-cloud-starter-alicloud-oss</artifactId>
24.            <version>2.2.0.RELEASE</version>
25.        </dependency>
26.        <dependency>
27.            <groupId>com.fasterxml.jackson.module</groupId>
28.            <artifactId>jackson-module-kotlin</artifactId>
29.        </dependency>
30.        <dependency>
31.            <groupId>org.jetbrains.kotlin</groupId>
32.            <artifactId>kotlin-reflect</artifactId>
33.        </dependency>
34.        <dependency>
35.            <groupId>org.jetbrains.kotlin</groupId>
36.            <artifactId>kotlin-stdlib-jdk8</artifactId>
37.        </dependency>
38.        <dependency>
39.            <groupId>org.jetbrains.kotlinx</groupId>
40.            <artifactId>kotlinx-coroutines-core</artifactId>
41.            <version>1.3.2</version>
42.        </dependency>
43.        <dependency>
44.            <groupId>org.springframework.boot</groupId>
45.            <artifactId>spring-boot-starter-test</artifactId>
46.            <version>2.2.1.RELEASE</version>
47.            <scope>test</scope>
48.        </dependency>
49.    </dependencies>
50.    <build>
51.        <sourceDirectory>${project.basedir}/src/main/kotlin</sourceDirectory>
52.        <testSourceDirectory>${project.basedir}/src/test/kotlin</testSourceDirectory>
53.        <plugins>
54.            <plugin>
```

```xml
55.            <groupId>org.springframework.boot</groupId>
56.            <artifactId>spring-boot-maven-plugin</artifactId>
57.        </plugin>
58.        <plugin>
59.            <groupId>org.jetbrains.kotlin</groupId>
60.            <artifactId>kotlin-maven-plugin</artifactId>
61.            <configuration>
62.                <args>
63.                    <arg>-Xjsr305=strict</arg>
64.                </args>
65.                <compilerPlugins>
66.                    <plugin>spring</plugin>
67.                    <plugin>jpa</plugin>
68.                </compilerPlugins>
69.            </configuration>
70.            <dependencies>
71.                <dependency>
72.                    <groupId>org.jetbrains.kotlin</groupId>
73.                    <artifactId>kotlin-maven-allopen</artifactId>
74.                    <version>${kotlin.version}</version>
75.                </dependency>
76.                <dependency>
77.                    <groupId>org.jetbrains.kotlin</groupId>
78.                    <artifactId>kotlin-maven-noarg</artifactId>
79.                    <version>${kotlin.version}</version>
80.                </dependency>
81.            </dependencies>
82.        </plugin>
83.    </plugins>
84. </build>
85.</project>
```

application.yml 文件如下所示,其中定义了 OSS 的访问地址和密钥信息,将 access-key 和 secret-key 替换为自己的即可:

```
1. server:
2.   port: 8082   #应用端口号
3. spring:
```

```yaml
4.    application:
5.      name: oss-app    #应用名
6.    cloud:
7.      alicloud:
8.        oss:
9.          endpoint: http://oss-cn-beijing.aliyuncs.com   #OSS 服务地址
10.         access-key:    #阿里云 access key
11.         secret-key:    #阿里云 secret key
```

OssApplication.kt 是一个启动类，启动了一个 Spring Boot 应用：

```kotlin
1. @SpringBootApplication
2. class OssApplication
3. // 启动函数
4. fun main(args: Array<String>) {
5.     runApplication<OssApplication>(*args)
6. }
```

OssController.kt 定义了四个接口，用于测试向 OSS 上传、下载文件。"/upload" 和 "/download" 使用 ossClient 上传及下载 oss-test.json 文件。"upload2" 和 "/file-resource" 使用 Resource 方式上传及下载 oss-test.json 文件。ossClient 通常用于操作大量文件对象的场景，如果只需读取少量文件，可以用 Resource 的形式得到文件对象。代码如下：

```kotlin
1.  @RestController
2.  class OssController {
3.      @Autowired
4.      lateinit var ossClient: OSS
5.      // 本地文件位置
6.      @Value("classpath:/oss-test.json")
7.      lateinit var localFile: Resource
8.      // 远程文件位置
9.      @Value("oss://kcglobal/test/oss-test.json")
10.     lateinit var remoteFile: Resource
11.     // 测试接口，上传文件到 OSS
12.     @GetMapping("/upload")
13.     fun upload(): String {
14.         try {
15.             val stream = this.javaClass.classLoader.getResourceAsStream("oss-test.json")
```

```kotlin
16.            ossClient.putObject("kcglobal", "test/oss-test.json", stream)
17.        } catch (e: Exception) {
18.            return "upload fail: " + e.message
19.        }
20.        return "upload success"
21.    }
22.    // 测试接口，从 OSS 读取文件
23.    @GetMapping("/file-resource")
24.    fun fileResource(): String {
25.        return try {
26.            "get file resource success. content: " +
27.                StreamUtils.copyToString(remoteFile.inputStream,
    Charset.forName(CharEncoding.UTF_8))
28.        } catch (e: Exception) {
29.            "get resource fail: " + e.message
30.        }
31.    }
32.    // 测试接口，从 OSS 读取文件
33.    @GetMapping("/download")
34.    fun download(): String {
35.        return try {
36.            val ossObject = ossClient.getObject("kcglobal", "test/oss-test.json")
37.            "download success, content: " +
    IOUtils.readStreamAsString(ossObject.objectContent, CharEncoding.UTF_8)
38.        } catch (e: Exception) {
39.            "download fail: " + e.message
40.        }
41.    }
42.    // 测试接口，上传文件到 OSS
43.    @GetMapping("/upload2")
44.    fun uploadWithOutputStream(): String {
45.        try {
46.            (this.remoteFile as WritableResource)
47.                .outputStream
48.                .use { outputStream ->
49.                    localFile.inputStream.use { inputStream ->
    StreamUtils.copy(inputStream, outputStream)
```

```kotlin
50.                    }
51.                }
52.            } catch (ex: Exception) {
53.                return "upload with outputStream failed"
54.            }
55.            return "upload success"
56.        }
57. }
```

OssImageController.kt 定义了两个接口，用于测试上传和下载图片。"image" 接口用于获取图片的访问地址，该地址有过期时间，"image/upload" 接口用于上传图片：

```kotlin
1. @RestController
2. class OssImageController {
3.     @Autowired
4.     lateinit var ossClient: OSS
5. 
6.     val imageExpireTime = 10 * 365 * 24 * 60 * 60 * 1000L;
7.     // 获取图片的访问地址
8.     @GetMapping("/image/{name}")
9.     fun getImageUrl(@PathVariable name: String): String? {
10.        val expiration = Date(Date().getTime() + imageExpireTime)
11.        val url = ossClient.generatePresignedUrl("kcglobal", "test/" + name, expiration)
12.        return url?.toString()
13.    }
14.    // 上传图片
15.    @GetMapping("/image/upload/{name}")
16.    fun uploadImage(@PathVariable name: String): String {
17.        var ret = ""
18.        var stream: InputStream ?= null
19.        try {
20.            val objectMetadata = ObjectMetadata()
21.            stream = this.javaClass.classLoader.getResourceAsStream("images/$name")
22.            objectMetadata.contentLength = stream.available().toLong()
23.            objectMetadata.cacheControl = "no-cache";
24.            objectMetadata.setHeader("Pragma", "no-cache");
```

```
25.            objectMetadata.contentType = "image/jpg"
26.            objectMetadata.contentDisposition = "inline;filename=$name"
27.            val putObject = ossClient.putObject("kcglobal", "test/$name", stream,
    objectMetadata)
28.            ret = putObject.eTag
29.        } catch (e: IOException) {
30.            println("upload file to oss error name=$name")
31.        } finally {
32.            try{
33.                if(stream != null) {
34.                    stream.close()
35.                }
36.            } catch (e: Exception) {
37.                e.printStackTrace()
38.            }
39.        }
40.        return ret
41.    }
42. }
```

8.4 分布式任务调度

SchedulerX 是阿里巴巴集团中间件团队开发的一款高性能、分布式的任务调度产品，在阿里巴巴内部有着广泛的使用，经过集团内上千个业务应用、历经多年打磨而成。截至 2016 年 6 月，在 SchedulerX 上每天平稳运行集团内的几十万个任务，完成每天几亿次的任务调度。本节介绍 Kotlin 集成 SchedulerX 的 SDK 进行定时任务开发的相关知识。

8.4.1 SchedulerX 介绍

SchedulerX 1.0 版本让任务实现了分布式。SchedulerX 1.0 提供了自主运维管理的后台，让用户能通过页面来配置、修改和管理定时任务，用户只需在页面上修改时间表达式，不需要重新发布运行定时任务的业务应用。SchedulerX 1.0 还能管理任务执行的生命周期，从任务执行开始一直到任务执行结束都有记录，用户能看到每次任务执行的开始时间和结束时间，还能看到执行成功或者失败，SchedulerX 1.0 还会为用户保留过去的执行记录，使用户可以查看定时任务历史执行记录。

此外，SchedulerX 1.0 能把一个执行耗时很长的定时任务拆分成多个子任务分片，分发到多台机器上并行执行，大大减少了定时任务的执行时间。

SchedulerX 2.0 版本提供了完善的任务调度体系。SchedulerX 2.0（DTS）进一步提升了用户体验，除优化编程模型、减少用户配置和程序接口之外，还新增了多项功能。SchedulerX 2.0（DTS）支持七种功能。

简单 job 单机版是每次随机选择一台机器执行任务，即任务只运行在一台机器上。但是为了防止单点故障还得解决多机备份的问题，当一台机器宕机的时候可以自动切换到其他正常工作的机器上去。

简单 job 广播版则是每次选择所有机器同时执行任务，比如需要定时更新本地内存的场景，这时就需要每台机器同时更新内存。

并行计算 job 是将一个耗时很长的大任务拆分成多个小的子任务然后分发到多台机器去并行执行。

图示计算（任务依赖），这个功能多用于在有业务数据依赖的多个任务之间按照严格先后顺序执行的场景。比如两个任务 A、B，其中 A 执行结束之后 B 才能开始执行。

脚本 job 是指 shell、PHP、Python 等定时执行的脚本任务，用户只需在 SchedulerX 2.0（DTS）管理后台配置要定时执行的 shell、php、python 等命令即可，用户不需要额外写任何代码。

SchedulerX 2.0（DTS）的管理运维控制台提供任务配置管理，以及历史执行记录查询，还有完善的监控报警功能。任务没有准时执行会给用户发送报警，任务执行超过预期的时间也会给用户发送报警，任务执行失败了也会给用户发送短信报警。

SchedulerX 2.0（DTS）还支持基于 SchedulerX 2.0 的二次开发，用户可以通过 SDK 中的 API 来创建、修改和删除任务。

SchedulerX 2.0（DTS）还支持超大规模定时触发器，用户可以通过 API 创建千亿量级的一次性定时触发器，比如每条交易订单创建的时候就在 SchedulerX 2.0 中创建一个定时触发器，用户通过设置这个触发器的触发时间，使得每到触发时间，事件通知交易系统就会提醒用户确认收货超时。

8.4.2　Kotlin 集成 SchedulerX

新建一个 Maven 子工程：chapter08-schedulerx，pom.xml 如下：

```xml
1.  <?xml version="1.0" encoding="UTF-8"?>
2.  <project xmlns="http://maven.apache.org/POM/4.0.0"
3.           xmlns:xsi="http://www.w3.org/2001/XMLSchema-instance"
4.           xsi:schemaLocation="http://maven.apache.org/POM/4.0.0
    http://maven.apache.org/xsd/maven-4.0.0.xsd">
5.      <parent>
6.          <artifactId>kotlinspringboot</artifactId>
7.          <groupId>io.kang.kotlinspringboot</groupId>
8.          <version>0.0.1-SNAPSHOT</version>
9.      </parent>
10.     <modelVersion>4.0.0</modelVersion>
11.     <!-- 子工程名-->
12.     <artifactId>chapter08-schedulerx</artifactId>
13.     <dependencies>
14.         <!-- Spring Cloud SchedulerX 依赖包-->
15.         <dependency>
16.             <groupId>com.alibaba.cloud</groupId>
17.             <artifactId>spring-cloud-starter-alicloud-schedulerx</artifactId>
18.             <version>2.2.0.RELEASE</version>
19.         </dependency>
20.         <dependency>
21.             <groupId>com.fasterxml.jackson.module</groupId>
22.             <artifactId>jackson-module-kotlin</artifactId>
23.         </dependency>
24.         <dependency>
25.             <groupId>org.jetbrains.kotlin</groupId>
26.             <artifactId>kotlin-reflect</artifactId>
27.         </dependency>
28.         <dependency>
29.             <groupId>org.jetbrains.kotlin</groupId>
30.             <artifactId>kotlin-stdlib-jdk8</artifactId>
31.         </dependency>
32.         <dependency>
33.             <groupId>org.jetbrains.kotlinx</groupId>
34.             <artifactId>kotlinx-coroutines-core</artifactId>
35.             <version>1.3.2</version>
36.         </dependency>
```

```xml
37.     </dependencies>
38.     <build>
39.         <sourceDirectory>${project.basedir}/src/main/kotlin</sourceDirectory>
40.         <testSourceDirectory>${project.basedir}/src/test/kotlin</testSourceDirectory>
41.         <plugins>
42.             <plugin>
43.                 <groupId>org.springframework.boot</groupId>
44.                 <artifactId>spring-boot-maven-plugin</artifactId>
45.             </plugin>
46.             <plugin>
47.                 <groupId>org.jetbrains.kotlin</groupId>
48.                 <artifactId>kotlin-maven-plugin</artifactId>
49.                 <configuration>
50.                     <args>
51.                         <arg>-Xjsr305=strict</arg>
52.                     </args>
53.                     <compilerPlugins>
54.                         <plugin>spring</plugin>
55.                         <plugin>jpa</plugin>
56.                     </compilerPlugins>
57.                 </configuration>
58.                 <dependencies>
59.                     <dependency>
60.                         <groupId>org.jetbrains.kotlin</groupId>
61.                         <artifactId>kotlin-maven-allopen</artifactId>
62.                         <version>${kotlin.version}</version>
63.                     </dependency>
64.                     <dependency>
65.                         <groupId>org.jetbrains.kotlin</groupId>
66.                         <artifactId>kotlin-maven-noarg</artifactId>
67.                         <version>${kotlin.version}</version>
68.                     </dependency>
69.                 </dependencies>
70.             </plugin>
71.         </plugins>
72.     </build>
73. </project>
```

application.yml 文件如下：

```yaml
1.  server:
2.    port: 8084  #应用端口号
3.  spring:
4.    application:
5.      name: SCX-APP  #应用名称
6.    cloud:
7.      alicloud:
8.        scx:
9.          group-id: ***  #SchedulerX 的 group id
10.       edas:
11.         namespace: cn-test  #namespace 名称
```

SpringBootApplication.kt 是一个 Spring Boot 启动类：

```kotlin
1. @SpringBootApplication
2. class ScxApplication
3. // 启动函数
4. fun main(args: Array<String>) {
5.     runApplication<ScxApplication>(*args)
6. }
```

SimpleTask.kt 定义了一个 job 处理类，打印 "Hello World"：

```kotlin
1. // job 处理类
2. class SimpleTask: ScxSimpleJobProcessor {
3.     override fun process(context: ScxSimpleJobContext?): ProcessResult {
4.         println("----------Hello world---------------")
5.         return ProcessResult(true)
6.     }
7. }
```

进入阿里云的 SchedulerX 任务列表页面，选择上方的"测试"区域，单击右上角的"新建 Job"，创建一个 job，如下所示：

1. job 分组：测试—***-*-*-****
2. job 处理接口：io.kang.schedule.SimpleTask

3. 类型：简单 job 单机版
4. 定时表达式：默认选项——0 * * * * ?
5. job 描述：无
6. 自定义参数：无

以上任务类型选择了"简单 job 单机版"，并且制定了 Cron 表达式为"0 * * * * ?"，这意味着，每过 1 分钟，任务将会被执行且只执行 1 次。

8.5 分布式事务

分布式事务是跨数据库的事务问题。为了解决数据库访问的瓶颈问题，分库是很常见的解决方案，不同用户可能落在不同的数据库里，原来在一个库里的事务操作，现在变成了跨数据库的事务操作。随着业务的不断增长，将业务不同模块中的服务拆分成微服务后，每个微服务都对应一个独立的数据源，数据源可能位于不同的机房，同时调用多个微服务很难保证同时成功，因此会产生跨服务分布式事务问题。本节介绍使用 Kotlin 集成 Seata 分布式框架进行分布式事务开发的相关知识。

8.5.1 分布式事务介绍

常见的分布式解决方案有：两阶段提交（2pc）、TCC 和事务消息。

两阶段提交协议涉及两种角色：一个是事务协调者（coordinator），负责协调多个参与者进行事务投票及提交（回滚）；另一个是多个事务参与者（participants），即本地事务执行者。处理步骤有以下两个。

投票阶段（voting phase）：协调者将通知事务参与者准备提交或取消事务，然后进入表决过程。参与者将告知协调者自己的决策——同意（事务参与者本地事务执行成功，但未提交）或取消（本地事务执行故障）。

提交阶段（commit phase）：收到参与者的通知后，协调者再向参与者发出通知，根据反馈情况决定各参与者提交还是回滚。

两阶段提交协议的实现很简单，但是需要数据库，需要 XA（eXtended Architecture）的强一致性支持，锁粒度大，性能差。

TCC 将事务提交分为 Try - Confirm - Cancel 三个操作。其和两阶段提交有点类似，Try 为第一阶段，Confirm - Cancel 为第二阶段，是一种应用层面侵入业务的两阶段提交。它不依赖 RM（资源管理器）对分布式事务的支持，而是通过对业务逻辑的分解来实现分布式事务。

事务消息更倾向于达成分布式事务的最终一致性，分布式事务的提交或回滚取决于事务发起方的业务需求，如 A 给 B 打了款并且成功了，那么下游业务 B 的钱款一定要增加这种场景，或某用户完成订单，用户积分一定要增加这种场景。

Seata 是阿里巴巴开源的分布式事务解决方案，对业务侵入度小，即减少技术架构上的微服务化所带来的分布式事务问题对业务的侵入；高性能，即减少分布式事务解决方案所带来的性能消耗。Seata 中有两种分布式事务实现方案，AT 及 TCC。AT 模式主要关注多数据库访问的数据一致性，当然也包括多服务下的多数据库数据访问的一致性问题。TCC 模式主要关注业务拆分，在横向扩展业务资源时，解决微服务间调用的一致性问题。

Seata AT 模式是基于 XA 事务演进而来的一个分布式事务中间件，XA 是一个基于数据库实现的分布式事务协议，本质上和两阶段提交一样，需要数据库支持，MySQL 5.6 以上版本支持 XA 协议，其他数据库如 Oracle、DB2 也实现了 XA 接口。Seata AT 模式中的各角色如图 8.6 所示。

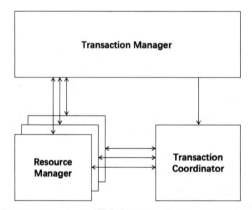

图8.6　Seata AT模式中的TC、RM、TM示意图

Transaction Coordinator（TC），事务协调器，维护全局事务的运行状态，负责协调并驱动全局事务的提交或回滚。Transaction Manager（TM），控制全局事务的边界，负责开启一个全局事务，并最终发起全局提交或全局回滚的决议。Resource Manager（RM），控制分支事务，负责分支注册、状态汇报，并接收事务协调器的指令及驱动分支（本地）事务的提交和回滚。

第一阶段，Seata 的 JDBC 数据源代理通过对业务 SQL 的解析，把业务数据更新前后的数据镜像组织成回滚日志，利用本地事务的 ACID 特性，将业务数据的更新和回滚日志写入同一个本地事务中并提交。这样，可以保证提交的任何业务数据的更新一定有相应的回滚日志存在。基于这样的机制，分支的本地事务便可以在全局事务的第一阶段提交，并马上释放本地事务锁定的资源。Seata 和 XA 事务的不同之处是，两阶段提交往往对资源的锁定需要持续到第二阶段实际的提交或者回滚操作，而有了回滚日志之后，可以在第一阶段释放对资源的锁定，降低了锁定范围，提高了效率，即使第二阶段发生异常需要回滚，只需找到 undolog 中对应的数据并反解析成 SQL 即可达到回滚的目的。同时，Seata 通过代理数据源将业务 SQL 的执行解析成 undolog 来与业务数据的更新同时入库，达到了对业务无侵入的效果。

　　第二阶段，如果决议是全局提交，此时分支事务已经完成提交，不需要同步协调处理（只需异步清理回滚日志），第二阶段可以非常快速地完成。如果决议是全局回滚，RM 收到协调器发来的回滚请求，通过 XID 和 Branch ID 找到相应的回滚日志记录，通过回滚记录生成反向的更新 SQL 并执行，以完成分支的回滚操作。

　　Seata 还针对 TCC 做了适配兼容，支持 TCC 事务方案，使用侵入业务上的补偿及事务管理器的协调来达到全局事务的一起提交及回滚。

8.5.2　Kotlin 集成 Seata

　　新建一个 Maven 子工程：chapter08-seata，这是一个服务提供者。pom 文件如下：

```xml
1.  <?xml version="1.0" encoding="UTF-8"?>
2.  <project xmlns="http://maven.apache.org/POM/4.0.0"
3.           xmlns:xsi="http://www.w3.org/2001/XMLSchema-instance"
4.           xsi:schemaLocation="http://maven.apache.org/POM/4.0.0 http://maven.apache.org/xsd/maven-4.0.0.xsd">
5.      <parent>
6.          <artifactId>kotlinspringboot</artifactId>
7.          <groupId>io.kang.kotlinspringboot</groupId>
8.          <version>0.0.1-SNAPSHOT</version>
9.      </parent>
10.     <modelVersion>4.0.0</modelVersion>
11.     <!-- 子工程名-->
12.     <artifactId>chapter08-seata</artifactId>
13.     <dependencies>
14.         <dependency>
15.             <groupId>com.alibaba.cloud</groupId>
```

```xml
16.         <artifactId>spring-cloud-starter-alibaba-nacos-discovery</artifactId>
17.         <version>2.1.1.RELEASE</version>
18.     </dependency>
19.     <!-- Spring Cloud Seata 依赖包-->
20.     <dependency>
21.         <groupId>com.alibaba.cloud</groupId>
22.         <artifactId>spring-cloud-starter-alibaba-seata</artifactId>
23.         <version>2.1.1.RELEASE</version>
24.     </dependency>
25.     <!-- Spring Boot Web 依赖包-->
26.     <dependency>
27.         <groupId>org.springframework.boot</groupId>
28.         <artifactId>spring-boot-starter-web</artifactId>
29.         <version>2.1.1.RELEASE</version>
30.     </dependency>
31.     <!-- Lombok 依赖包-->
32.     <dependency>
33.         <groupId>org.projectlombok</groupId>
34.         <artifactId>lombok</artifactId>
35.     </dependency>
36.     <!-- MyBatis 依赖包-->
37.     <dependency>
38.         <groupId>org.mybatis.spring.boot</groupId>
39.         <artifactId>mybatis-spring-boot-starter</artifactId>
40.         <version>2.1.1</version>
41.     </dependency>
42.     <!-- MySQL 驱动-->
43.     <dependency>
44.         <groupId>mysql</groupId>
45.         <artifactId>mysql-connector-java</artifactId>
46.     </dependency>
47.     <dependency>
48.         <groupId>com.fasterxml.jackson.module</groupId>
49.         <artifactId>jackson-module-kotlin</artifactId>
50.     </dependency>
51.     <dependency>
52.         <groupId>org.jetbrains.kotlin</groupId>
```

```xml
53.        <artifactId>kotlin-reflect</artifactId>
54.    </dependency>
55.    <dependency>
56.        <groupId>org.jetbrains.kotlin</groupId>
57.        <artifactId>kotlin-stdlib-jdk8</artifactId>
58.    </dependency>
59.    <dependency>
60.        <groupId>org.jetbrains.kotlinx</groupId>
61.        <artifactId>kotlinx-coroutines-core</artifactId>
62.        <version>1.3.2</version>
63.    </dependency>
64.    <dependency>
65.        <groupId>org.springframework.boot</groupId>
66.        <artifactId>spring-boot-starter-test</artifactId>
67.        <scope>test</scope>
68.    </dependency>
69. </dependencies>
70. <build>
71.    <sourceDirectory>${project.basedir}/src/main/kotlin</sourceDirectory>
72.    <testSourceDirectory>${project.basedir}/src/test/kotlin</testSourceDirectory>
73.    <plugins>
74.        <plugin>
75.            <groupId>org.springframework.boot</groupId>
76.            <artifactId>spring-boot-maven-plugin</artifactId>
77.        </plugin>
78.        <plugin>
79.            <groupId>org.jetbrains.kotlin</groupId>
80.            <artifactId>kotlin-maven-plugin</artifactId>
81.            <configuration>
82.                <args>
83.                    <arg>-Xjsr305=strict</arg>
84.                </args>
85.                <compilerPlugins>
86.                    <plugin>spring</plugin>
87.                    <plugin>jpa</plugin>
88.                </compilerPlugins>
89.            </configuration>
```

```xml
90.            <dependencies>
91.                <dependency>
92.                    <groupId>org.jetbrains.kotlin</groupId>
93.                    <artifactId>kotlin-maven-allopen</artifactId>
94.                    <version>${kotlin.version}</version>
95.                </dependency>
96.                <dependency>
97.                    <groupId>org.jetbrains.kotlin</groupId>
98.                    <artifactId>kotlin-maven-noarg</artifactId>
99.                    <version>${kotlin.version}</version>
100.               </dependency>
101.           </dependencies>
102.       </plugin>
103.    </plugins>
104. </build>
105.</project>
```

application.yml 定义如下，定义了数据库连接信息，Nacos 服务注册中心地址，Seata 事务服务组：

```yaml
1.  server:
2.    port: 8085   #应用端口号
3.  spring:
4.    application:
5.      name: sca-provider   #应用名
6.    datasource:
7.      driver-class-name: com.mysql.jdbc.Driver   #数据库驱动
8.      url: jdbc:mysql://127.0.0.1:3306/video?characterEncoding=utf-8&serverTimezone=UTC   #数据库连接地址
9.      username: root   #数据库连接用户名
10.     password: 123456   #数据库连接密码
11.   cloud:
12.     nacos:
13.       discovery:
14.         server-addr: 127.0.0.1:8848   #Nacos 注册中心地址
15.     alibaba:
16.       seata:
17.         tx-service-group: sca-provider-group   #事务服务组名称
```

```
18. mybatis:
19.   mapper-locations: classpath:mapper/*.xml    #数据库操作的相关 XML 文件所在位置
```

registry.conf 定义了 Seata 配置信息存储方式,可以存储在本地文件、Nacos、Eureka、Redis、Zookeeper、Etcd、Sofa 中,这里用本地文件方式存储配置信息:

```
1.  registry {
2.    # file、Nacos、Eureka、Redis、Zookeeper、Consul、Etcd3、Sofa
3.    # 采用文件方式,文件名是 file.conf
4.    type = "file"
5.    file {
6.      name = "file.conf"
7.    }
8.  }
9.  config {
10.   # file、Nacos、Apollo、Zookeeper、Consul、Etcd3、Spring Cloud Config
11.   # 采用文件方式,文件名是 file.conf
12.   type = "file"
13.   file {
14.     name = "file.conf"
15.   }
16. }
```

file.conf 定义了 Seata 的配置信息,vgroup_mapping 的 sca-provider-group 和 application.yml 定义的 tx-service-group 保持一致,disableGlobalTransaction 用于控制是否开启全局事务:

```
1.  transport {
2.    # 采用 TCP 通信协议
3.    type = "TCP"
4.    # 服务端使用 NIO
5.    server = "NIO"
6.    # 开启心跳
7.    heartbeat = true
8.    # 允许客户端批量发送请求
9.    enableClientBatchSendRequest = true
10.   # Netty 线程工厂配置
11.   threadFactory {
12.     bossThreadPrefix = "NettyBoss"
```

```
13.      workerThreadPrefix = "NettyServerNIOWorker"
14.      serverExecutorThread-prefix = "NettyServerBizHandler"
15.      shareBossWorker = false
16.      clientSelectorThreadPrefix = "NettyClientSelector"
17.      clientSelectorThreadSize = 1
18.      clientWorkerThreadPrefix = "NettyClientWorkerThread"
19.      # Netty boss 线程大小
20.      bossThreadSize = 1
21.      # 工作线程大小
22.      workerThreadSize = "default"
23.    }
24.    shutdown {
25.      # 服务销毁后，等待 3 秒
26.      wait = 3
27.    }
28.    # 序列化方式
29.    serialization = "seata"
30.    # 不压缩
31.    compressor = "none"
32.  }
33.  service {
34.    # 事务服务组映射
35.    vgroup_mapping.sca-provider-group = "default"
36.    # 当 registry.type=file 时，设置该属性，不要设置多个地址
37.    default.grouplist = "127.0.0.1:8091"
38.    # 降级，目前不支持
39.    enableDegrade = false
40.    # 开启 Seata 事务
41.    disableGlobalTransaction = false
42.  }
43.  client {
44.    # 资源管理相关配置
45.    rm {
46.      asyncCommitBufferLimit = 10000
47.      lock {
48.        retryInterval = 10
49.        retryTimes = 30
```

```
50.        retryPolicyBranchRollbackOnConflict = true
51.      }
52.      reportRetryCount = 5
53.      tableMetaCheckEnable = false
54.      reportSuccessEnable = false
55.    }
56.    # 事务管理相关配置
57.    tm {
58.      commitRetryCount = 5
59.      rollbackRetryCount = 5
60.    }
61.    undo {
62.      dataValidation = true
63.      logSerialization = "jackson"
64.      logTable = "undo_log"
65.    }
66.    log {
67.      exceptionRate = 100
68.    }
69.  }
70.  support {
71.    # 不允许自动代理数据源
72.    spring.datasource.autoproxy = false
73.  }
```

ProviderApp.kt 是启动类，启动了一个 Spring Boot 服务，排除了 Spring Boot 自带的数据源配置类，采用自定义的数据源配置类：

```
1.  // 排除数据源自动配置类，开启 Eureka 客户端注解
2.  @SpringBootApplication(exclude = [DataSourceAutoConfiguration::class])
3.  @EnableDiscoveryClient
4.  @MapperScan("io.kang.provider.mapper")
5.  class ProviderApp
6.  // 启动函数
7.  fun main(args: Array<String>) {
8.      runApplication<ProviderApp>(*args)
9.  }
```

DataSourceConfiguration.kt 定义了 Seata 事务使用的数据源，数据库连接池用 Druid 加载 MyBatis 的 mapper 和配置：

```kotlin
1.  @Configuration
2.  @EnableConfigurationProperties(MybatisProperties::class)
3.  class DataSourceConfiguration {
4.      // 初始化数据源
5.      @Bean
6.      @ConfigurationProperties(prefix = "spring.datasource")
7.      fun dataSource(): DataSource {
8.          return DruidDataSource()
9.      }
10.     // 初始化数据源代理
11.     @Bean
12.     @Primary
13.     fun dataSourceProxy(dataSource: DataSource): DataSourceProxy {
14.         return DataSourceProxy(dataSource)
15.     }
16.     // 初始化 sqlSessionFactory
17.     @Bean
18.     fun sqlSessionFactoryBean(dataSourceProxy: DataSourceProxy,
19.                 mybatisProperties: MybatisProperties): SqlSessionFactoryBean {
20.         val bean = SqlSessionFactoryBean()
21.         bean.setDataSource(dataSourceProxy)
22.         val resolver = PathMatchingResourcePatternResolver()
23.         try {
24.             val locations = resolver.getResources(mybatisProperties.mapperLocations[0])
25.             bean.setMapperLocations(*locations)
26.             if (StringUtils.isNotBlank(mybatisProperties.configLocation)) {
27.                 val resources = resolver.getResources(mybatisProperties.configLocation)
28.                 bean.setConfigLocation(resources[0])
29.             }
30.         } catch (e: IOException) {
31.             e.printStackTrace()
32.         }
33.         return bean
34.     }
35. }
```

TbUserMapper.kt 定义了一个 mapper 接口，用于向 tb_user 表插入记录：

```
1.  // mapper 接口
2.  interface TbUserMapper {
3.      fun insert(record: TbUser): Int
4.  }
```

TbUser.kt 定义了一个 TbUser 实体：

```
1.  // 实体类
2.  @Data
3.  class TbUser : Serializable {
4.      var id: Int? = null
5.      var name: String? = null
6.      var age: Int? = null
7.  }
```

ProviderController.kt 提供了一个接口 "/add/user"，向数据库插入一条 TbUser 记录：

```
1.  @RestController
2.  class ProviderController {
3.      @Autowired
4.      lateinit var userMapper: TbUserMapper
5.      // 测试接口
6.      @PostMapping("/add/user")
7.      fun add(@RequestBody user: TbUser) {
8.          println("add user: $user")
9.          user.name = "provider"
10.         userMapper.insert(user)
11.     }
12. }
```

新建一个 Maven 子工程：chapter08-seata-consumer，这是一个消费方，调用 chapter08-seata 定义的 "/add/user" 接口。pom 文件如下：

```
1.  <?xml version="1.0" encoding="UTF-8"?>
2.  <project xmlns="http://maven.apache.org/POM/4.0.0"
3.           xmlns:xsi="http://www.w3.org/2001/XMLSchema-instance"
```

```xml
4.         xsi:schemaLocation="http://maven.apache.org/POM/4.0.0
   http://maven.apache.org/xsd/maven-4.0.0.xsd">
5.     <parent>
6.         <artifactId>kotlinspringboot</artifactId>
7.         <groupId>io.kang.kotlinspringboot</groupId>
8.         <version>0.0.1-SNAPSHOT</version>
9.     </parent>
10.    <modelVersion>4.0.0</modelVersion>
11.    <!-- 子工程名 -->
12.    <artifactId>chapter08-seata-consumer</artifactId>
13.    <dependencies>
14.        <dependency>
15.            <groupId>com.alibaba.cloud</groupId>
16.            <artifactId>spring-cloud-starter-alibaba-nacos-discovery</artifactId>
17.            <version>2.1.1.RELEASE</version>
18.        </dependency>
19.        <!-- Spring Cloud Seata 依赖包 -->
20.        <dependency>
21.            <groupId>com.alibaba.cloud</groupId>
22.            <artifactId>spring-cloud-starter-alibaba-seata</artifactId>
23.            <version>2.1.1.RELEASE</version>
24.        </dependency>
25.        <!-- Spring Boot Actuator 依赖包 -->
26.        <dependency>
27.            <groupId>org.springframework.boot</groupId>
28.            <artifactId>spring-boot-actuator</artifactId>
29.            <version>2.2.2.RELEASE</version>
30.        </dependency>
31.        <!-- Spring Boot Web 依赖包 -->
32.        <dependency>
33.            <groupId>org.springframework.boot</groupId>
34.            <artifactId>spring-boot-starter-web</artifactId>
35.        </dependency>
36.        <!-- Spring Cloud Feign 依赖包 -->
37.        <dependency>
38.            <groupId>org.springframework.cloud</groupId>
```

```xml
39.            <artifactId>spring-cloud-starter-openfeign</artifactId>
40.            <version>2.2.2.RELEASE</version>
41.        </dependency>
42.        <!-- Lombok 依赖包 -->
43.        <dependency>
44.            <groupId>org.projectlombok</groupId>
45.            <artifactId>lombok</artifactId>
46.        </dependency>
47.        <!-- MyBatis 依赖包 -->
48.        <dependency>
49.            <groupId>org.mybatis.spring.boot</groupId>
50.            <artifactId>mybatis-spring-boot-starter</artifactId>
51.            <version>2.1.1</version>
52.        </dependency>
53.        <!-- 数据库连接依赖包 -->
54.        <dependency>
55.            <groupId>mysql</groupId>
56.            <artifactId>mysql-connector-java</artifactId>
57.        </dependency>
58.        <dependency>
59.            <groupId>com.fasterxml.jackson.module</groupId>
60.            <artifactId>jackson-module-kotlin</artifactId>
61.        </dependency>
62.        <dependency>
63.            <groupId>org.jetbrains.kotlin</groupId>
64.            <artifactId>kotlin-reflect</artifactId>
65.        </dependency>
66.        <dependency>
67.            <groupId>org.jetbrains.kotlin</groupId>
68.            <artifactId>kotlin-stdlib-jdk8</artifactId>
69.        </dependency>
70.        <dependency>
71.            <groupId>org.jetbrains.kotlinx</groupId>
72.            <artifactId>kotlinx-coroutines-core</artifactId>
73.            <version>1.3.2</version>
74.        </dependency>
75.    </dependencies>
```

```xml
76. <build>
77.     <sourceDirectory>${project.basedir}/src/main/kotlin</sourceDirectory>
78.     <testSourceDirectory>${project.basedir}/src/test/kotlin</testSourceDirectory>
79.     <plugins>
80.         <plugin>
81.             <groupId>org.springframework.boot</groupId>
82.             <artifactId>spring-boot-maven-plugin</artifactId>
83.         </plugin>
84.         <plugin>
85.             <groupId>org.jetbrains.kotlin</groupId>
86.             <artifactId>kotlin-maven-plugin</artifactId>
87.             <configuration>
88.                 <args>
89.                     <arg>-Xjsr305=strict</arg>
90.                 </args>
91.                 <compilerPlugins>
92.                     <plugin>spring</plugin>
93.                     <plugin>jpa</plugin>
94.                 </compilerPlugins>
95.             </configuration>
96.             <dependencies>
97.                 <dependency>
98.                     <groupId>org.jetbrains.kotlin</groupId>
99.                     <artifactId>kotlin-maven-allopen</artifactId>
100.                    <version>${kotlin.version}</version>
101.                </dependency>
102.                <dependency>
103.                    <groupId>org.jetbrains.kotlin</groupId>
104.                    <artifactId>kotlin-maven-noarg</artifactId>
105.                    <version>${kotlin.version}</version>
106.                </dependency>
107.            </dependencies>
108.        </plugin>
109.    </plugins>
110. </build>
111. </project>
```

application.yml 定义了数据库配置、Nacos 服务注册中心地址、Seata 事务服务组：

```yaml
1.  server:
2.    port: 8086       #应用端口号
3.  spring:
4.    datasource:
5.      driver-class-name: com.mysql.jdbc.Driver   #数据库驱动类
6.      url: jdbc:mysql://127.0.0.1:3306/video?characterEncoding=utf-8&serverTimezone=UTC   #数据库连接地址
7.      username: root   #数据库连接用户名
8.      password: 123456      #数据库连接密码
9.    cloud:
10.     nacos:
11.       discovery:
12.         server-addr: 127.0.0.1:8848   #Nacos 服务注册中心地址
13.     alibaba:
14.       seata:
15.         # Seata 事务组名称，对应 file.conf 文件中的 vgroup_mapping.sca-customer-seata-tx-service-group
16.         tx-service-group: sca-customer-group
17.     application:
18.       name: sca-customer    #应用名称
19. mybatis:
20.   mapper-locations: classpath:mapper/*.xml   #数据库操作相关的 XML 文件所在的位置
```

file.conf、registry.conf 和 chapter08-seata 的定义基本一致，区别在于，file.conf 的如下配置项，vgroup_mapping 的 sca-customer-group 和 application.yml 中定义的 seata-tx-service-group 保持一致：

```
1. service {
2.   #事务服务组映射
3.   vgroup_mapping.sca-customer-group = "default"
4. }
```

CustomerApp.kt 定义了一个启动类，排除 Spring Boot 自带的数据源配置类：

```
1. // 扫描 @FeignClient 注解
```

```
2.  @SpringBootApplication(exclude = [DataSourceAutoConfiguration::class])
3.  @EnableDiscoveryClient
4.  @EnableFeignClients
5.  @MapperScan("io.kang.consumer.mapper")
6.  class CustomerApp
7.  // 启动函数
8.  fun main(args: Array<String>) {
9.      runApplication<CustomerApp>(*args)
10. }
```

ProviderFeignService.kt 通过 Feign 方式访问"chapter08-seata"定义的"/add/user"接口：

```
1.  //Feign 接口
2.  @FeignClient(value = "sca-provider")
3.  interface ProviderFeignService {
4.      @PostMapping("/add/user")
5.      fun add(@RequestBody user: TbUser)
6.  }
```

UserController.kt 定义了一个接口"/seata/user/add"，通过调用"/user/add"接口，用本地方法 localSave 向 tb_user 表插入一条 TbUser 记录。使用@GlobalTransactional 开启 Seata 分布式事务，当调用"/user/add"接口后抛出异常，chapter08-seata 的"/user/add"接口回滚，本地方法 localSave 也回滚，数据库中没有任何记录。chapter08-seata 的"/user/add"接口、本地方法 localSave 分别向 tb_user 插入记录，由于是分布式事务，出现异常可回滚，避免了数据不一致。

将配置文件 file.conf 中的 disableGlobalTransaction 设置为 true，关闭分布式事务，再次调用"seata/user/add"，tb_user 表中出现两条记录，表示事务没有回滚。

```
1.  @RestController
2.  class UserController {
3.      @Autowired
4.      lateinit var userMapper: TbUserMapper
5.
6.      @Autowired
7.      lateinit var providerFeignService: ProviderFeignService
8.      // 测试接口，测试分布式事务
9.      @PostMapping("/seata/user/add")
```

```kotlin
10.     @GlobalTransactional(rollbackFor = [Exception::class]) // 开启全局事务
11.     fun add(@RequestBody user: TbUser) {
12.         println("globalTransactional begin, Xid: ${RootContext.getXID()}")
13.         // 使用本地方法保存 user
14.         localSave(user)
15.         // 远程调用接口保存 user
16.         providerFeignService.add(user)
17.         // 抛出异常，测试 Seata 事务
18.         throw RuntimeException()
19.     }
20.     private fun localSave(user: TbUser) {
21.         user.name = "customer"
22.         userMapper.insert(user)
23.     }
24. }
```

图 8.7 和图 8.8 所示的两个模块都出现了回滚日志："Branch Rollbacked result: PhaseTwo_Rollbacked"。

```
load [io.seata.rm.datasource.undo.parser.ProtostuffUndoLogParser] class fail. io/protostuff/runtime/Delegate
load UndoLogParser[jackson] extension by class[io.seata.rm.datasource.undo.parser.JacksonUndoLogParser]
Flipping property: sca-provider.ribbon.ActiveConnectionsLimit to use NEXT property: niws.loadbalancer.availabilityFilteringRule.activeConnectionsLimit = 21474
Client: sca-provider instantiated a LoadBalancer: DynamicServerListLoadBalancer:{NFLoadBalancer:name=sca-provider,current list of Servers=[],Load balancer sta
Using serverListUpdater PollingServerListUpdater
Flipping property: sca-provider.ribbon.ActiveConnectionsLimit to use NEXT property: niws.loadbalancer.availabilityFilteringRule.activeConnectionsLimit = 21474
DynamicServerListLoadBalancer for client sca-provider initialized: DynamicServerListLoadBalancer:{NFLoadBalancer:name=sca-provider,current list of Servers=[19
tion failure:0;    Total blackout seconds:0;   Last connection made:Thu Jan 01 08:00:00 CST 1970;    First connection made: Thu Jan 01 08:00:00 CST 1970;     Act
onMessage:xid=192.168.126.1:8091:2045610353,branchId=2045610356,branchType=AT,resourceId=jdbc:mysql://127.0.0.1:3306/video,applicationData=null
Branch Rollbacking: 192.168.126.1:8091:2045610353 2045610356 jdbc:mysql://127.0.0.1:3306/video
Could not found property transaction.undo.data.validation, try to use default value instead.
Flipping property: sca-provider.ribbon.ActiveConnectionsLimit to use NEXT property: niws.loadbalancer.availabilityFilteringRule.activeConnectionsLimit = 21474
xid 192.168.126.1:8091:2045610353 branch 2045610356, undo_log deleted with GlobalFinished
Branch Rollbacked result: PhaseTwo_Rollbacked
[192.168.126.1:8091:2045610353] rollback status:Rollbacked
Servlet.service() for servlet [dispatcherServlet] in context with path [] threw exception [Request processing failed; nested exception is java.lang.RuntimeExc
```

图8.7 chapter08-seata-consumer回滚日志

```
Could not found property client.report.retry.count, try to use default value instead.
Could not found property client.lock.retry.policy.branch-rollback-on-conflict, try to use default value instead.
Could not found property client.lock.retry.internal, try to use default value instead.
Could not found property client.lock.retry.times, try to use default value instead.
load LoadBalance[null] extension by class[io.seata.discovery.loadbalance.RandomLoadBalance]
Could not found property transaction.undo.log.table, try to use default value instead.
Could not found property transaction.undo.log.serialization, try to use default value instead.
load [io.seata.rm.datasource.undo.parser.ProtostuffUndoLogParser] class fail. io/protostuff/runtime/RuntimeEnv
load UndoLogParser[jackson] extension by class[io.seata.rm.datasource.undo.parser.JacksonUndoLogParser]
onMessage:xid=192.168.126.1:8091:2045610353,branchId=2045610362,branchType=AT,resourceId=jdbc:mysql://127.0.0.1:3306/video,applicationData=null
Branch Rollbacking: 192.168.126.1:8091:2045610353 2045610362 jdbc:mysql://127.0.0.1:3306/video
Could not found property transaction.undo.data.validation, try to use default value instead.
xid 192.168.126.1:8091:2045610353 branch 2045610362, undo_log deleted with GlobalFinished
Branch Rollbacked result: PhaseTwo_Rollbacked
```

图8.8 chapter08-seata回滚日志

8.6　Spring Cloud Dubbo

Dubbo 是阿里巴巴内部的 SOA 服务化治理方案的核心框架，每天为 2000 多个服务提供多于 30 亿次的访问支持，并被广泛应用于阿里巴巴集团的各成员站点。Dubbo 自 2011 年开源后，已被许多非阿里系公司使用，目前是 Apache 的顶级项目。本节介绍使用 Kotlin 集成 Spring Cloud Dubbo 进行微服务开发的方法。

8.6.1　Dubbo 介绍

Dubbo 在国内拥有巨大的用户群，大家希望在使用 Dubbo 的同时享受 Spring Cloud 的生态系统，因而出现了各种各样的整合方案。但是因为服务中心不同，各种整合方案并不是那么自然，直到 Spring Cloud Alibaba 这个项目出现，由官方提供了 Nacos 服务注册中心，才将这个问题完美解决，并且提供了 Dubbo 和 Spring Cloud 整合的方案 Dubbo Spring Cloud。

Dubbo Spring Cloud 构建在原生的 Spring Cloud 之上，其在服务治理方面的能力可被认为是 Spring Cloud Plus，不仅完全覆盖了 Spring Cloud 的原生特性，而且提供了更为稳定和成熟的实现。其与 Spring Cloud 的特性对比如表 8.1 所示。

表 8.1　Spring Cloud、Dubbo Spring Cloud 特性对比

功能组件	Spring Cloud	Dubbo Spring Cloud
分布式配置（Distributed configuration）	Git、Zookeeper、Consul、JDBC	Spring Cloud 分布式配置 + Dubbo 配置中心
服务注册与发现（Service registration and discovery）	Eureka、Zookeeper、Consul	Spring Cloud 原生注册中心 + Dubbo 原生注册中心
负载均衡（Load balancing）	Ribbon（随机、轮询等算法）	Dubbo 内建实现（随机、轮询等算法 + 权重等特性）
服务熔断（Circuit Breakers）	Spring Cloud Hystrix	Spring Cloud Hystrix + Alibaba Sentinel 等
服务调用（Service-to-service calls）	OpenFeign、RestTemplate	Spring Cloud 服务调用 + Dubbo @Reference
链路跟踪（Tracing）	Spring Cloud Sleuth + Zipkin	Zipkin、Opentracing 等

Dubbo Spring Cloud 的主要特性如下所述。

面向接口代理的高性能远程过程调用：提供了高性能的基于代理的远程调用能力，服

务以接口为粒度,屏蔽了远程调用底层细节。

智能负载均衡:内置多种负载均衡策略,可智能感知下游节点的健康状况,显著减少调用延迟,提高系统吞吐量。

服务自动注册与发现:支持多种注册中心服务,服务实例上下线实时感知。

高度可扩展能力:遵循"微内核+插件"的设计原则,所有核心能力如 Protocol、Transport、Serialization 被设计为扩展点,平等对待内置实现和第三方实现。

运行期流量调度,内置条件、脚本等路由策略:通过配置不同的路由规则,可轻松实现灰度发布及同机房优先等功能。

可视化的服务治理与运维:提供丰富的服务治理、运维工具,随时查询服务元数据、服务健康状态及调用统计,实时下发路由策略、调整配置参数。

8.6.2　Kotlin 集成 Spring Cloud Dubbo

定义一个 Maven 子工程:chapter08-dubbo,这是一个 Dubbo 服务提供方。pom 文件如下:

```xml
1.  <?xml version="1.0" encoding="UTF-8"?>
2.  <project xmlns="http://maven.apache.org/POM/4.0.0"
3.           xmlns:xsi="http://www.w3.org/2001/XMLSchema-instance"
4.           xsi:schemaLocation="http://maven.apache.org/POM/4.0.0
    http://maven.apache.org/xsd/maven-4.0.0.xsd">
5.      <parent>
6.          <artifactId>kotlinspringboot</artifactId>
7.          <groupId>io.kang.kotlinspringboot</groupId>
8.          <version>0.0.1-SNAPSHOT</version>
9.      </parent>
10.     <modelVersion>4.0.0</modelVersion>
11.     <!-- 子工程名-->
12.     <artifactId>chapter08-dubbo</artifactId>
13.     <dependencies>
14.         <!-- Spring Boot Actuator 依赖包-->
15.         <dependency>
16.             <groupId>org.springframework.boot</groupId>
17.             <artifactId>spring-boot-actuator</artifactId>
```

```xml
18.        </dependency>
19.        <!-- Spring Boot Web 依赖包-->
20.        <dependency>
21.            <groupId>org.springframework.boot</groupId>
22.            <artifactId>spring-boot-starter-web</artifactId>
23.        </dependency>
24.        <!-- Spring Cloud Dubbo 依赖包-->
25.        <dependency>
26.            <groupId>com.alibaba.cloud</groupId>
27.            <artifactId>spring-cloud-starter-dubbo</artifactId>
28.            <version>2.2.0.RELEASE</version>
29.        </dependency>
30.        <!-- Spring Cloud Nacos 依赖包-->
31.        <dependency>
32.            <groupId>com.alibaba.cloud</groupId>
33.            <artifactId>spring-cloud-starter-alibaba-nacos-discovery</artifactId>
34.            <version>2.2.0.RELEASE</version>
35.        </dependency>
36.        <dependency>
37.            <groupId>com.fasterxml.jackson.module</groupId>
38.            <artifactId>jackson-module-kotlin</artifactId>
39.        </dependency>
40.        <dependency>
41.            <groupId>org.jetbrains.kotlin</groupId>
42.            <artifactId>kotlin-reflect</artifactId>
43.        </dependency>
44.        <dependency>
45.            <groupId>org.jetbrains.kotlin</groupId>
46.            <artifactId>kotlin-stdlib-jdk8</artifactId>
47.        </dependency>
48.        <dependency>
49.            <groupId>org.jetbrains.kotlinx</groupId>
50.            <artifactId>kotlinx-coroutines-core</artifactId>
51.            <version>1.3.2</version>
52.        </dependency>
53.        <dependency>
54.            <groupId>org.springframework.boot</groupId>
```

```xml
55.            <artifactId>spring-boot-starter-test</artifactId>
56.            <scope>test</scope>
57.        </dependency>
58.    </dependencies>
59.    <build>
60.        <sourceDirectory>${project.basedir}/src/main/kotlin</sourceDirectory>
61.        <testSourceDirectory>${project.basedir}/src/test/kotlin</testSourceDirectory>
62.        <plugins>
63.            <plugin>
64.                <groupId>org.springframework.boot</groupId>
65.                <artifactId>spring-boot-maven-plugin</artifactId>
66.            </plugin>
67.            <plugin>
68.                <groupId>org.jetbrains.kotlin</groupId>
69.                <artifactId>kotlin-maven-plugin</artifactId>
70.                <configuration>
71.                    <args>
72.                        <arg>-Xjsr305=strict</arg>
73.                    </args>
74.                    <compilerPlugins>
75.                        <plugin>spring</plugin>
76.                        <plugin>jpa</plugin>
77.                    </compilerPlugins>
78.                </configuration>
79.                <dependencies>
80.                    <dependency>
81.                        <groupId>org.jetbrains.kotlin</groupId>
82.                        <artifactId>kotlin-maven-allopen</artifactId>
83.                        <version>${kotlin.version}</version>
84.                    </dependency>
85.                    <dependency>
86.                        <groupId>org.jetbrains.kotlin</groupId>
87.                        <artifactId>kotlin-maven-noarg</artifactId>
88.                        <version>${kotlin.version}</version>
89.                    </dependency>
90.                </dependencies>
91.            </plugin>
```

```
92.         </plugins>
93.     </build>
94. </project>
```

application.yml 文件的定义如下，定义了 Dubbo 接口所在的包路径，使用的协议、端口及 Nacos 注册中心地址：

```
1.  server:
2.    port: 8087   #应用端口号
3.  dubbo:
4.    scan:
5.      # Dubbo 服务扫描基准包
6.      base-packages: io.kang.provider.dubbo
7.    protocol:
8.      # Dubbo 协议
9.      name: dubbo
10.     # Dubbo 协议端口（-1 表示自增端口，从 20880 开始）
11.     port: -1
12.   registry:
13.     # 挂载到 Nacos 注册中心
14.     address: nacos://127.0.0.1:8848
15.   cloud:
16.     subscribed-services: ""   #订阅的 Dubbo 服务
17. spring:
18.   cloud:
19.     nacos:
20.       discovery:
21.         server-addr: 127.0.0.1:8848   #Nacos 注册中心地址
22.     application:
23.       name: dubbo-provider   #应用名称
```

DubboProviderApp.kt 是启动类，开启了服务注册注解：

```
1. @SpringBootApplication
2. @EnableDiscoveryClient
3. class DubboProviderApp
4. // 启动函数
5. fun main(args: Array<String>) {
```

```
6.     runApplication<DubboProviderApp>(*args)
7. }
```

DubboEchoService.kt 定义了一个对外提供的 Dubbo 服务接口:

```
1. // 服务接口
2. interface DubboEchoService {
3.     fun echo(name: String): String
4. }
```

DubboEchoServiceImpl.kt 实现了 DubboEchoService 接口:

```
1. // 实现接口
2. @Service
3. class DubboEchoServiceImpl: DubboEchoService {
4.     override fun echo(name: String): String {
5.         return "DubboEchoServiceImpl#echo hi $name"
6.     }
7. }
```

新建一个 Maven 子工程:chapter08-dubbo-consumer,这是一个服务消费方。pom 文件如下:

```
1.  <?xml version="1.0" encoding="UTF-8"?>
2.  <project xmlns="http://maven.apache.org/POM/4.0.0"
3.           xmlns:xsi="http://www.w3.org/2001/XMLSchema-instance"
4.           xsi:schemaLocation="http://maven.apache.org/POM/4.0.0
    http://maven.apache.org/xsd/maven-4.0.0.xsd">
5.      <parent>
6.          <artifactId>kotlinspringboot</artifactId>
7.          <groupId>io.kang.kotlinspringboot</groupId>
8.          <version>0.0.1-SNAPSHOT</version>
9.      </parent>
10.     <modelVersion>4.0.0</modelVersion>
11.     <!-- 子工程名-->
12.     <artifactId>chapter08-dubbo-consumer</artifactId>
13.     <dependencies>
14.         <!-- 依赖 chapter08-dubbo 子工程 -->
```

```xml
15.    <dependency>
16.        <artifactId>chapter08-dubbo</artifactId>
17.        <groupId>io.kang.kotlinspringboot</groupId>
18.        <version>0.0.1-SNAPSHOT</version>
19.    </dependency>
20.    <!-- Spring Boot Actuator 依赖包 -->
21.    <dependency>
22.        <groupId>org.springframework.boot</groupId>
23.        <artifactId>spring-boot-actuator</artifactId>
24.    </dependency>
25.    <!-- Spring Boot Web 依赖包 -->
26.    <dependency>
27.        <groupId>org.springframework.boot</groupId>
28.        <artifactId>spring-boot-starter-web</artifactId>
29.    </dependency>
30.    <!-- Spring Cloud Dubbo 依赖包 -->
31.    <dependency>
32.        <groupId>com.alibaba.cloud</groupId>
33.        <artifactId>spring-cloud-starter-dubbo</artifactId>
34.        <version>2.2.0.RELEASE</version>
35.    </dependency>
36.    <!-- Spring Cloud Nacos 依赖包 -->
37.    <dependency>
38.        <groupId>com.alibaba.cloud</groupId>
39.        <artifactId>spring-cloud-starter-alibaba-nacos-discovery</artifactId>
40.        <version>2.2.0.RELEASE</version>
41.    </dependency>
42.    <dependency>
43.        <groupId>com.fasterxml.jackson.module</groupId>
44.        <artifactId>jackson-module-kotlin</artifactId>
45.    </dependency>
46.    <dependency>
47.        <groupId>org.jetbrains.kotlin</groupId>
48.        <artifactId>kotlin-reflect</artifactId>
49.    </dependency>
50.    <dependency>
51.        <groupId>org.jetbrains.kotlin</groupId>
```

```xml
52.         <artifactId>kotlin-stdlib-jdk8</artifactId>
53.     </dependency>
54.     <dependency>
55.         <groupId>org.jetbrains.kotlinx</groupId>
56.         <artifactId>kotlinx-coroutines-core</artifactId>
57.         <version>1.3.2</version>
58.     </dependency>
59.     <dependency>
60.         <groupId>org.springframework.boot</groupId>
61.         <artifactId>spring-boot-starter-test</artifactId>
62.         <scope>test</scope>
63.     </dependency>
64. </dependencies>
65. <build>
66.     <sourceDirectory>${project.basedir}/src/main/kotlin</sourceDirectory>
67.     <testSourceDirectory>${project.basedir}/src/test/kotlin</testSourceDirectory>
68.     <plugins>
69.         <plugin>
70.             <groupId>org.springframework.boot</groupId>
71.             <artifactId>spring-boot-maven-plugin</artifactId>
72.         </plugin>
73.         <plugin>
74.             <groupId>org.jetbrains.kotlin</groupId>
75.             <artifactId>kotlin-maven-plugin</artifactId>
76.             <configuration>
77.                 <args>
78.                     <arg>-Xjsr305=strict</arg>
79.                 </args>
80.                 <compilerPlugins>
81.                     <plugin>spring</plugin>
82.                     <plugin>jpa</plugin>
83.                 </compilerPlugins>
84.             </configuration>
85.             <dependencies>
86.                 <dependency>
87.                     <groupId>org.jetbrains.kotlin</groupId>
88.                     <artifactId>kotlin-maven-allopen</artifactId>
```

```xml
89.                <version>${kotlin.version}</version>
90.            </dependency>
91.            <dependency>
92.                <groupId>org.jetbrains.kotlin</groupId>
93.                <artifactId>kotlin-maven-noarg</artifactId>
94.                <version>${kotlin.version}</version>
95.            </dependency>
96.        </dependencies>
97.      </plugin>
98.    </plugins>
99.  </build>
100.</project>
```

application.yml 文件中定义了 Nacos 注册中心地址和订阅的 Dubbo 服务：

```yaml
1. server:
2.   port: 8088   #服务端口号
3. dubbo:
4.   registry:
5.     # 挂载到 Spring Cloud 注册中心
6.     address: nacos://127.0.0.1:8848
7.   cloud:
8.     subscribed-services: dubbo-provider   #订阅 dubbo-provider 服务
9.   consumer:
10.    check: false   #关闭所有服务的启动时检查
11. spring:
12.   application:
13.     name: dubbo-consumer   #应用名称
14.   cloud:
15.     nacos:
16.       discovery:
17.         server-addr: 127.0.0.1:8848   #Nacos 服务注册中心地址
```

DubboConsumerApp.kt 是启动类，开启了服务注册注解：

```kotlin
1. @SpringBootApplication
2. @EnableDiscoveryClient
3. class DubboConsumerApp
```

```kotlin
4.    // 启动函数
5.    fun main(args: Array<String>) {
6.        runApplication<DubboConsumerApp>(*args)
7.    }
```

CustomerController.kt 定义了一个接口"/dubbo/echo"，通过@Reference 注解访问 Dubbo 服务——dubboEchoService：

```kotlin
1.  @RestController
2.  class CustomerController {
3.      // 引用 dubboEchoService
4.      @Reference
5.      lateinit var dubboEchoService: DubboEchoService
6.      // 测试接口，通过 Dubbo 访问 dubboEchoService
7.      @GetMapping("/dubbo/echo/{name}")
8.      fun dubboEcho(@PathVariable("name") name: String): String {
9.          return dubboEchoService.echo(name)
10.     }
11. }
```

8.7 小结

本章介绍了 Spring Cloud Alibaba 生态系统中常用的组件，包括服务注册和发现、服务限流降级、消息驱动、阿里对象云存储、分布式任务调度、分布式事务、Dubbo 等。通过了解这些组件，你可以在实际的项目中选择性地集成并使用它们，以提高开发效率。

第 9 章
Kotlin 集成服务监控和服务链路监控

　　微服务系统的分布式特点使得系统监控比较困难,本章将介绍服务监控中间件 Prometheus、Grafana 和服务链路监控中间件 Zipkin、SkyWalking,通过示例展示 Kotlin 集成服务监控、服务链路监控的方法。

9.1　Prometheus、Grafana 介绍

　　Prometheus 是由 SoundCloud 开源的监控告警解决方案。Prometheus 存储的是时序数据,即按相同时序（相同名称和标签）,以时间维度存储的连续数据的集合。时序（time series）是由名字（metric）以及一组 key/value 标签定义的,具有相同名字以及标签的数据属于相同时序。metric 有 Counter、Gauge、Histogram、Summary 四种类型。Counter 是一种累加的 metric,如请求的个数、结束的任务数、出现的错误数等。Gauge 是常规的 metric,如温度,可任意加减,与时间没有关系,是可以任意变化的数据。Histogram 是柱状图,用于观察结果采样、分组及统计。Summary 类似 Histogram,用于表示一段时间内数据的采样结果。

　　Prometheus 提供数据查询 DSL 语言——PromQL。PromQL 支持通过名称及标签进行查询,如 http_requests_total 等价于 {name="http_requests_total"}。查询条件支持正则匹配,如 http_requests_total{code!="200"} 表示查询 code 不为"200"的数据。支持内置函数,如将浮点数转换为整数,查看每秒数据。支持模糊查询、比较查询、范围查询、聚合、统计等。

查询结果的类型有：瞬时数据（Instant vector），包含一组时序，每个时序只有一个点，例如，http_requests_total；区间数据（Range vector），包含一组时序，每个时序有多个点，例如，http_requests_total[5m]；纯量数据（Scalar），纯量只有一个数字，没有时序，例如，count(http_requests_total)。

Prometheus 基于 HTTP 采用 pull（拉取）方式收集数据。由于各个被监控的对象不在一个子网或防火墙内导致无法直接拉取各 target 数据，需要将不同数据汇总到推送网关（Pushgateway），再由 Prometheus 统一收集。Prometheus 向 Altermanager 推送报警，架构如图 9.1 所示。

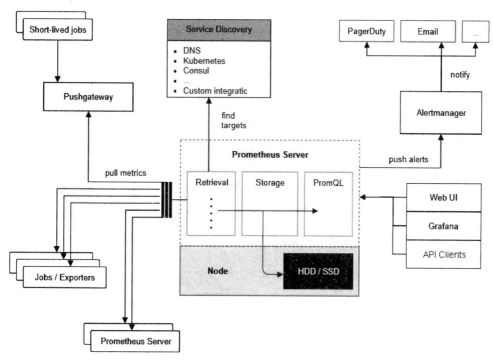

图9.1　Prometheus架构图

Grafana 是一款用 Go 语言开发的开源数据可视化工具，可用于数据监控和数据统计，带有告警功能。目前使用 Grafana 的用户很多。Grafana 具有如下特点。

- **可视化**：快速且灵活的客户端具有多种选项，面板插件用不同的方式可视化指标和日志。
- **报警**：可视化地为重要的指标定义警报规则，Grafana 将持续评估它们，并发送通知。

- **通知**：警报状态更改时，Grafana 会发出通知。它还可以接收电子邮件通知。
- **动态仪表盘**：使用模板变量创建动态和可重用的仪表盘，这些模板变量作为下拉菜单出现在仪表盘顶部。
- **混合数据源**：在同一张图中混合不同的数据源。可以根据每个查询指定数据源，这甚至适用于自定义数据源。
- **注释**：注释来自不同的数据源图表。将鼠标指针悬停在事件上可以显示完整的事件元数据和标记。
- **过滤器**：允许用户动态创建新的"键-值"过滤器，这些过滤器将自动应用于使用该数据源的所有查询。

Grafana 从 Prometheus 获取监控指标，并进行可视化展示，如图 9.2 所示。

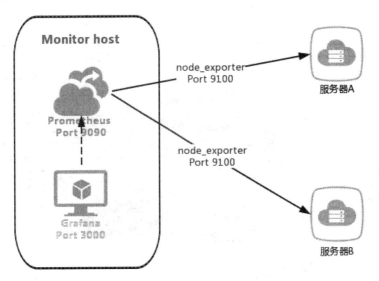

图9.2　Prometheus、Grafana监控服务示意图

9.2　Kotlin 集成 Prometheus、Grafana

新建一个 Maven 子工程：chapter09-prometheus，pom 文件如下：

```
1.  <?xml version="1.0" encoding="UTF-8"?>
2.  <project xmlns="http://maven.apache.org/POM/4.0.0"
3.           xmlns:xsi="http://www.w3.org/2001/XMLSchema-instance"
```

```xml
4.        xsi:schemaLocation="http://maven.apache.org/POM/4.0.0
   http://maven.apache.org/xsd/maven-4.0.0.xsd">
5.    <parent>
6.        <artifactId>kotlinspringboot</artifactId>
7.        <groupId>io.kang.kotlinspringboot</groupId>
8.        <version>0.0.1-SNAPSHOT</version>
9.    </parent>
10.   <modelVersion>4.0.0</modelVersion>
11.   <!-- 子工程名-->
12.   <artifactId>chapter09-prometheus</artifactId>
13.   <dependencies>
14.       <!-- Spring Boot Web 依赖包-->
15.       <dependency>
16.           <groupId>org.springframework.boot</groupId>
17.           <artifactId>spring-boot-starter-web</artifactId>
18.           <version>2.2.1.RELEASE</version>
19.       </dependency>
20.       <!-- Spring Boot Actuator 依赖包-->
21.       <dependency>
22.           <groupId>org.springframework.boot</groupId>
23.           <artifactId>spring-boot-starter-actuator</artifactId>
24.           <version>2.2.1.RELEASE</version>
25.       </dependency>
26.       <!-- Micrometer 指标依赖包-->
27.       <dependency>
28.           <groupId>io.micrometer</groupId>
29.           <artifactId>micrometer-registry-prometheus</artifactId>
30.           <version>1.3.5</version>
31.       </dependency>
32.       <dependency>
33.           <groupId>com.fasterxml.jackson.module</groupId>
34.           <artifactId>jackson-module-kotlin</artifactId>
35.       </dependency>
36.       <dependency>
37.           <groupId>org.jetbrains.kotlin</groupId>
38.           <artifactId>kotlin-reflect</artifactId>
```

```xml
39.        </dependency>
40.        <dependency>
41.            <groupId>org.jetbrains.kotlin</groupId>
42.            <artifactId>kotlin-stdlib-jdk8</artifactId>
43.        </dependency>
44.        <dependency>
45.            <groupId>org.jetbrains.kotlinx</groupId>
46.            <artifactId>kotlinx-coroutines-core</artifactId>
47.            <version>1.3.2</version>
48.        </dependency>
49.    </dependencies>
50.    <build>
51.        <sourceDirectory>${project.basedir}/src/main/kotlin</sourceDirectory>
52.        <testSourceDirectory>${project.basedir}/src/test/kotlin</testSourceDirectory>
53.        <plugins>
54.            <plugin>
55.                <groupId>org.springframework.boot</groupId>
56.                <artifactId>spring-boot-maven-plugin</artifactId>
57.            </plugin>
58.            <plugin>
59.                <groupId>org.jetbrains.kotlin</groupId>
60.                <artifactId>kotlin-maven-plugin</artifactId>
61.                <configuration>
62.                    <args>
63.                        <arg>-Xjsr305=strict</arg>
64.                    </args>
65.                    <compilerPlugins>
66.                        <plugin>spring</plugin>
67.                        <plugin>jpa</plugin>
68.                    </compilerPlugins>
69.                </configuration>
70.                <dependencies>
71.                    <dependency>
72.                        <groupId>org.jetbrains.kotlin</groupId>
73.                        <artifactId>kotlin-maven-allopen</artifactId>
74.                        <version>${kotlin.version}</version>
```

```xml
75.            </dependency>
76.            <dependency>
77.                <groupId>org.jetbrains.kotlin</groupId>
78.                <artifactId>kotlin-maven-noarg</artifactId>
79.                <version>${kotlin.version}</version>
80.            </dependency>
81.            </dependencies>
82.        </plugin>
83.      </plugins>
84.    </build>
85. </project>
```

application.yml 文件的定义如下，暴露了监控接口，Prometheus 通过这些接口采集监控信息：

```yaml
1.  server:
2.    port: 8090   # 应用端口号
3.  spring:
4.    application:
5.      name: springboot-prometheus    # 应用名称
6.  management:
7.    endpoints:
8.      web:
9.        exposure:
10.          include: '*'   # 对外暴露的监控接口
11.    endpoint:
12.      prometheus:
13.        enabled: true   # 开启 Prometheus 监控
14.    metrics:
15.      export:
16.        prometheus:
17.          enabled: true   # 输出 Prometheus 监控指标
18.      tags:
19.        application: ${spring.application.name}   # 监控的标签
```

PrometheusApp.kt 定义了启动类，configurer 方法对外提供监控信息，这些监控信息使用 spring.application.name 作为公共标签：

```kotlin
1.  @SpringBootApplication
2.  class PrometheusApp {
3.      // 配置 Micrometer 监控
4.      @Bean
5.      fun configurer(@Value("\${spring.application.name}") applicationName: String): MeterRegistryCustomizer<MeterRegistry> {
6.          return MeterRegistryCustomizer<MeterRegistry>
7.          { registry -> registry.config().commonTags("application", applicationName) }
8.      }
9.  }
10. // 启动函数
11. fun main(args: Array<String>) {
12.     runApplication<PrometheusApp>(*args)
13. }
```

IndexController.kt 提供了一个测试接口，新建一个指标，监控接口访问次数：

```kotlin
1.  @RestController
2.  class IndexController {
3.      private var counter: Counter? = null
4.      constructor(registry: MeterRegistry, @Value("\${spring.application.name}") applicationName: String) {
5.          this.counter = registry.counter("index.api.counter", "application", applicationName)
6.      }
7.      // 测试接口，每调用 1 次，index.api.counter 指标会加 1
8.      @GetMapping("/prometheus")
9.      fun testPrometheus(): String {
10.         this.counter?.increment()
11.         return "hello prometheus"
12.     }
13. }
```

启动 chapter09-prometheus 工程，并启动 Prometheus 和 Grafana，可以看到图 9.3 和图 9.4 所示的监控页面。采用 Micrometer 模板，可显示应用启动的时间、应用持续的时间及堆使用率等。

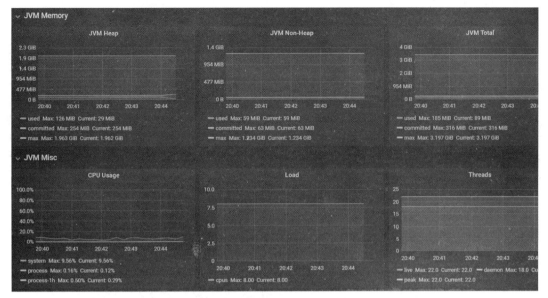

图9.3　JVM Memory指标

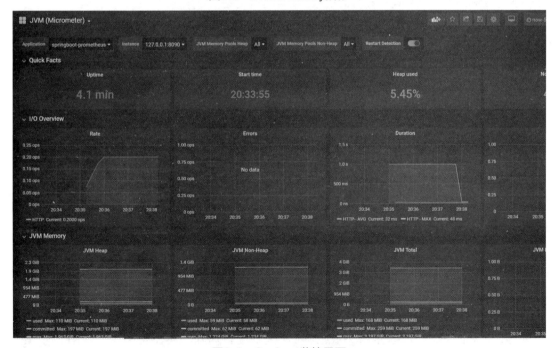

图9.4　Grafana监控界面

此外，还可以自定义监控指标，并显示在 Grafana 上。例如，对接口"/prometheus"

的监控，以及监控接口的访问次数，如图 9.5 所示。

图9.5　Grafana监控自定义指标

9.3　Kotlin 集成 Zipkin

　　Zipkin 是 Twitter 基于 Google 的分布式监控系统 Dapper（论文）的开源工具，Zipkin 用于跟踪分布式服务之间的应用数据链路、分析处理延时，可帮助我们改进系统的性能和定位故障。Span 是 Zipkin 的基本工作单元，一次链路调用就会创建一个 Span。Trace 是一组 Span 的集合，表示一条调用链路。例如，服务 A 调用服务 B 然后调用服务 C，这个 A→B→C 的链路就是一条 Trace，而每个服务，例如，B 就是一个 Span，如果在服务 B 中另起两个线程分别调用了 D、E，那么 D、E 就是 B 的子 Span。Zipkin 的架构如图 9.6 所示。

　　左上部分代表了客户端，分别为：Instrumented client，是使用了 Zipkin 客户端工具的服务调用方；Instrumented server，是使用了 Zipkin 客户端工具的服务提供方；Non-Instrumented server，是未使用 Trace 工具的服务提供方，当然还可能存在未使用工具的调用方。一个调用链路是贯穿 Instrumented client→Instrumented server 的，每经过一个服务都会以 Span 的形式通过 Transport 把经过自身的请求上报到 Zipkin 服务端中。

　　右边虚线框中的内容代表了 Zipkin 的服务端。UI 提供 Web 页面，用来展示 Zipkin 中的调用链和系统依赖关系等；Collector 对各个客户端暴露，负责接收调用数据，支持 HTTP、MQ 等；Storage 负责与各个存储适配后存储数据，支持内存、MySQL、ES 等；API 为 Web 界面提供查询存储中的数据的接口。

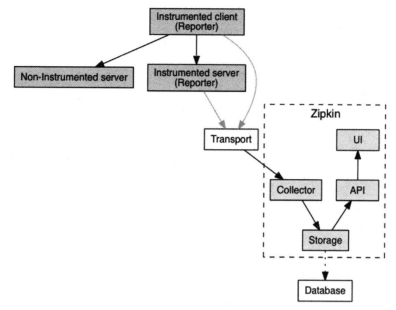

图9.6　Zipkin的架构图

新建 Maven 子工程 chapter10-zipkin1，这是一个服务提供方，pom 文件如下：

```
1.  <?xml version="1.0" encoding="UTF-8"?>
2.  <project xmlns="http://maven.apache.org/POM/4.0.0"
3.      xmlns:xsi="http://www.w3.org/2001/XMLSchema-instance"
4.      xsi:schemaLocation="http://maven.apache.org/POM/4.0.0
    http://maven.apache.org/xsd/maven-4.0.0.xsd">
5.      <parent>
6.          <artifactId>kotlinspringboot</artifactId>
7.          <groupId>io.kang.kotlinspringboot</groupId>
8.          <version>0.0.1-SNAPSHOT</version>
9.      </parent>
10.     <modelVersion>4.0.0</modelVersion>
11.     <!-- 子工程名-->
12.     <artifactId>chapter10-zipkin1</artifactId>
13.     <dependencies>
14.         <!-- Spring Boot Web 依赖包-->
15.         <dependency>
16.             <groupId>org.springframework.boot</groupId>
```

```xml
17.            <artifactId>spring-boot-starter-web</artifactId>
18.            <version>2.2.1.RELEASE</version>
19.        </dependency>
20.        <!-- Spring Cloud Eureka 客户端依赖包-->
21.        <dependency>
22.            <groupId>org.springframework.cloud</groupId>
23.            <artifactId>spring-cloud-starter-netflix-eureka-client</artifactId>
24.            <version>2.2.1.RELEASE</version>
25.        </dependency>
26.        <!-- Spring Cloud Feign 依赖包-->
27.        <dependency>
28.            <groupId>org.springframework.cloud</groupId>
29.            <artifactId>spring-cloud-starter-openfeign</artifactId>
30.            <version>2.2.1.RELEASE</version>
31.        </dependency>
32.        <!-- Spring Cloud Zipkin 依赖包-->
33.        <dependency>
34.            <groupId>org.springframework.cloud</groupId>
35.            <artifactId>spring-cloud-starter-zipkin</artifactId>
36.            <version>2.2.1.RELEASE</version>
37.        </dependency>
38.        <dependency>
39.            <groupId>com.fasterxml.jackson.module</groupId>
40.            <artifactId>jackson-module-kotlin</artifactId>
41.        </dependency>
42.        <dependency>
43.            <groupId>org.jetbrains.kotlin</groupId>
44.            <artifactId>kotlin-reflect</artifactId>
45.        </dependency>
46.        <dependency>
47.            <groupId>org.jetbrains.kotlin</groupId>
48.            <artifactId>kotlin-stdlib-jdk8</artifactId>
49.        </dependency>
50.        <dependency>
51.            <groupId>org.jetbrains.kotlinx</groupId>
52.            <artifactId>kotlinx-coroutines-core</artifactId>
```

```xml
53.        <version>1.3.2</version>
54.      </dependency>
55.    </dependencies>
56.    <build>
57.      <sourceDirectory>${project.basedir}/src/main/kotlin</sourceDirectory>
58.      <testSourceDirectory>${project.basedir}/src/test/kotlin</testSourceDirectory>
59.      <plugins>
60.        <plugin>
61.          <groupId>org.springframework.boot</groupId>
62.          <artifactId>spring-boot-maven-plugin</artifactId>
63.        </plugin>
64.        <plugin>
65.          <groupId>org.jetbrains.kotlin</groupId>
66.          <artifactId>kotlin-maven-plugin</artifactId>
67.          <configuration>
68.            <args>
69.              <arg>-Xjsr305=strict</arg>
70.            </args>
71.            <compilerPlugins>
72.              <plugin>spring</plugin>
73.              <plugin>jpa</plugin>
74.            </compilerPlugins>
75.          </configuration>
76.          <dependencies>
77.            <dependency>
78.              <groupId>org.jetbrains.kotlin</groupId>
79.              <artifactId>kotlin-maven-allopen</artifactId>
80.              <version>${kotlin.version}</version>
81.            </dependency>
82.            <dependency>
83.              <groupId>org.jetbrains.kotlin</groupId>
84.              <artifactId>kotlin-maven-noarg</artifactId>
85.              <version>${kotlin.version}</version>
86.            </dependency>
87.          </dependencies>
88.        </plugin>
```

```
89.         </plugins>
90.       </build>
91. </project>
```

application.yml 文件的内容如下：

```
1. server:
2.   port: 8102   #应用端口号
3. spring:
4.   application:
5.     name: zipkin-service2   #应用名
6.   zipkin:
7.     base-url: http://localhost:9411   #Zipkin 服务器地址
8. eureka:
9.   client:
10.    service-url:
11.      defaultZone: http://localhost:8761/eureka/   #服务注册中心地址
```

ZipkinApplication1.kt 启动了一个 Spring Boot 应用：

```
1. // 开启 Eureka Client 注解
2. @SpringBootApplication
3. @EnableEurekaClient
4. class ZipkinApplication1 {
5.     // 开启负载均衡
6.     @Bean
7.     @LoadBalanced
8.     fun restTemplate(): RestTemplate {
9.         return RestTemplate()
10.    }
11. }
12. // 启动函数
13. fun main(args: Array<String>) {
14.     runApplication<ZipkinApplication1>(*args)
15. }
```

ZipkinController.kt 定义了一个测试接口"service2"：

```
1.  @RestController
2.  class ZipkinController {
3.      // 测试接口
4.      @GetMapping("service2")
5.      fun service2(): String {
6.          return "Hello, I'm service2"
7.      }
8.  }
```

新建 Maven 子工程 chapter10-zipkin，这是一个服务消费方，pom 文件如下：

```
1.  <?xml version="1.0" encoding="UTF-8"?>
2.  <project xmlns="http://maven.apache.org/POM/4.0.0"
3.           xmlns:xsi="http://www.w3.org/2001/XMLSchema-instance"
4.           xsi:schemaLocation="http://maven.apache.org/POM/4.0.0
    http://maven.apache.org/xsd/maven-4.0.0.xsd">
5.      <parent>
6.          <artifactId>kotlinspringboot</artifactId>
7.          <groupId>io.kang.kotlinspringboot</groupId>
8.          <version>0.0.1-SNAPSHOT</version>
9.      </parent>
10.     <modelVersion>4.0.0</modelVersion>
11.     <!-- 子工程名-->
12.     <artifactId>chapter10-zipkin</artifactId>
13.     <dependencies>
14.         <!-- Spring Boot Web 依赖包-->
15.         <dependency>
16.             <groupId>org.springframework.boot</groupId>
17.             <artifactId>spring-boot-starter-web</artifactId>
18.             <version>2.2.1.RELEASE</version>
19.         </dependency>
20.         <!-- Spring Cloud Eureka 客户端依赖包-->
21.         <dependency>
22.             <groupId>org.springframework.cloud</groupId>
23.             <artifactId>spring-cloud-starter-netflix-eureka-client</artifactId>
24.             <version>2.2.1.RELEASE</version>
25.         </dependency>
```

```xml
26.        <!-- Spring Cloud Feign 依赖包-->
27.        <dependency>
28.            <groupId>org.springframework.cloud</groupId>
29.            <artifactId>spring-cloud-starter-openfeign</artifactId>
30.            <version>2.2.1.RELEASE</version>
31.        </dependency>
32.        <!-- Spring Cloud Zipkin 依赖包-->
33.        <dependency>
34.            <groupId>org.springframework.cloud</groupId>
35.            <artifactId>spring-cloud-starter-zipkin</artifactId>
36.            <version>2.2.1.RELEASE</version>
37.        </dependency>
38.        <dependency>
39.            <groupId>com.fasterxml.jackson.module</groupId>
40.            <artifactId>jackson-module-kotlin</artifactId>
41.        </dependency>
42.        <dependency>
43.            <groupId>org.jetbrains.kotlin</groupId>
44.            <artifactId>kotlin-reflect</artifactId>
45.        </dependency>
46.        <dependency>
47.            <groupId>org.jetbrains.kotlin</groupId>
48.            <artifactId>kotlin-stdlib-jdk8</artifactId>
49.        </dependency>
50.        <dependency>
51.            <groupId>org.jetbrains.kotlinx</groupId>
52.            <artifactId>kotlinx-coroutines-core</artifactId>
53.            <version>1.3.2</version>
54.        </dependency>
55.    </dependencies>
56.    <build>
57.        <sourceDirectory>${project.basedir}/src/main/kotlin</sourceDirectory>
58.        <testSourceDirectory>${project.basedir}/src/test/kotlin</testSourceDirectory>
59.        <plugins>
60.            <plugin>
61.                <groupId>org.springframework.boot</groupId>
62.                <artifactId>spring-boot-maven-plugin</artifactId>
```

```xml
63.            </plugin>
64.            <plugin>
65.                <groupId>org.jetbrains.kotlin</groupId>
66.                <artifactId>kotlin-maven-plugin</artifactId>
67.                <configuration>
68.                    <args>
69.                        <arg>-Xjsr305=strict</arg>
70.                    </args>
71.                    <compilerPlugins>
72.                        <plugin>spring</plugin>
73.                        <plugin>jpa</plugin>
74.                    </compilerPlugins>
75.                </configuration>
76.                <dependencies>
77.                    <dependency>
78.                        <groupId>org.jetbrains.kotlin</groupId>
79.                        <artifactId>kotlin-maven-allopen</artifactId>
80.                        <version>${kotlin.version}</version>
81.                    </dependency>
82.                    <dependency>
83.                        <groupId>org.jetbrains.kotlin</groupId>
84.                        <artifactId>kotlin-maven-noarg</artifactId>
85.                        <version>${kotlin.version}</version>
86.                    </dependency>
87.                </dependencies>
88.            </plugin>
89.        </plugins>
90.    </build>
91. </project>
```

application.yml 文件的内容如下：

```
1. server:
2.   port: 8101   # 应用端口号
3. spring:
4.   application:
5.     name: zipkin-service1   # 应用名
```

```yaml
6.  zipkin:
7.    base-url: http://localhost:9411  # Zipkin 服务地址
8.  eureka:
9.    client:
10.     service-url:
11.       defaultZone: http://localhost:8761/eureka/  # Eureka 注册中心地址
```

ZipkinApplication.kt 的定义如下，这是一个启动类，启动了一个 Spring Boot 服务：

```kotlin
1.  // 开启 Eureka Client 注解
2.  @SpringBootApplication
3.  @EnableEurekaClient
4.  class ZipkinApplication {
5.      @Bean
6.      @LoadBalanced
7.      fun restTemplate(): RestTemplate {
8.          return RestTemplate()
9.      }
10. }
11. // 启动函数
12. fun main(args: Array<String>) {
13.     runApplication<ZipkinApplication>(*args)
14. }
```

ZipkinController.kt，调用 zipkin-service2 提供的服务：

```kotlin
1.  @RestController
2.  class ZipkinController {
3.      @Autowired
4.      lateinit var restTemplate: RestTemplate
5.      // 测试接口，使用 Ribbon 方式调用/service2 接口
6.      @GetMapping("service1")
7.      fun service1(): String {
8.          return restTemplate.getForObject("http://zipkin-service2/service2", String::class)
9.      }
10. }
```

通过"service1"调用"service2"接口,在 ZipKin 监控界面可以看到调用信息,每次调用都有记录,如图 9.7 所示。

图9.7　Zipkin监控界面

具体到某次调用,可以看到调用每个服务接口的时间,如图 9.8 所示。

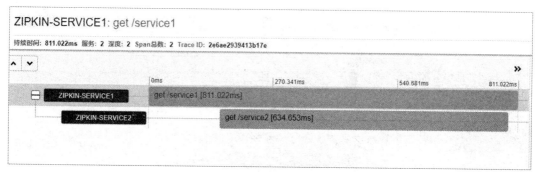

图9.8　Zipkin监控服务接口调用时间

此外,还可以展示服务的依赖关系,如图 9.9 所示。

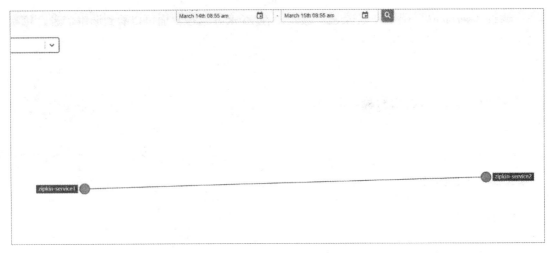

图9.9 服务依赖关系图

9.4 Kotlin 集成 SkyWalking

SkyWalking 是由国内开源爱好者吴晟开源并提交到 Apache 孵化器的产品，它同时吸收了 Zipkin/Pinpoint/CAT 的设计思路，支持非侵入式埋点。这是一款基于分布式跟踪的应用程序性能监控系统。SkyWalking 包括分布式追踪、性能指标分析、应用和服务依赖分析等，其架构如图 9.10 所示。

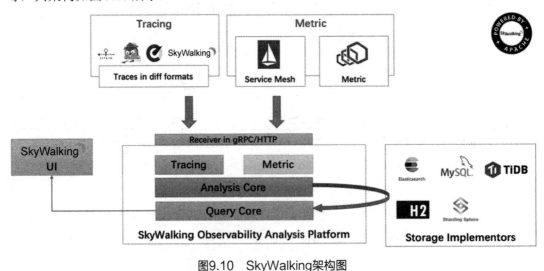

图9.10 SkyWalking架构图

SkyWalking 的核心是一个数据分析和度量结果的存储平台,通过 HTTP 或 gRPC 方式向 SkyWalking Collecter 提交分析和度量数据,SkyWalking Collecter 对数据进行分析和聚合,并将数据存储到 Elasticsearch、H2、MySQL、TiDB 等其一,最后可以通过 SkyWalking UI 的可视化界面对最终的结果进行呈现。SkyWalking 支持从多个来源和多种格式收集数据——多种语言的 SkyWalking Agent、Zipkin v1/v2、Istio 勘测、Envoy 度量等数据格式。

SkyWalking 收集各种格式的数据进行存储,然后进行展示。搭建 SkyWalking 服务需要关注的是 SkyWalking Collecter、SkyWalking UI 和存储设备。

SkyWalking 采用 Java 探针技术,在代码层面对应用程序没有任何侵入,使用起来简单方便,当然其具体实现需要针对不同的框架及服务提供探针插件。

新建 Maven 子工程 chapter10-skywalking1,这是一个服务提供方,pom 文件如下:

```xml
<?xml version="1.0" encoding="UTF-8"?>
<project xmlns="http://maven.apache.org/POM/4.0.0"
         xmlns:xsi="http://www.w3.org/2001/XMLSchema-instance"
         xsi:schemaLocation="http://maven.apache.org/POM/4.0.0
         http://maven.apache.org/xsd/maven-4.0.0.xsd">
    <parent>
        <artifactId>kotlinspringboot</artifactId>
        <groupId>io.kang.kotlinspringboot</groupId>
        <version>0.0.1-SNAPSHOT</version>
    </parent>
    <modelVersion>4.0.0</modelVersion>
    <!-- 子工程名-->
    <artifactId>chapter10-skywalking1</artifactId>
    <dependencies>
        <!-- Spring Boot Web 依赖包-->
        <dependency>
            <groupId>org.springframework.boot</groupId>
            <artifactId>spring-boot-starter-web</artifactId>
            <version>2.2.1.RELEASE</version>
        </dependency>
        <!-- Spring Boot Nacos 依赖包-->
        <dependency>
            <groupId>com.alibaba.cloud</groupId>
            <artifactId>spring-cloud-starter-alibaba-nacos-discovery</artifactId>
```

```xml
24.            <version>2.1.1.RELEASE</version>
25.        </dependency>
26.        <!-- Spring Boot Feign 依赖包-->
27.        <dependency>
28.            <groupId>org.springframework.cloud</groupId>
29.            <artifactId>spring-cloud-starter-openfeign</artifactId>
30.            <version>2.2.1.RELEASE</version>
31.        </dependency>
32.        <dependency>
33.            <groupId>com.fasterxml.jackson.module</groupId>
34.            <artifactId>jackson-module-kotlin</artifactId>
35.        </dependency>
36.        <dependency>
37.            <groupId>org.jetbrains.kotlin</groupId>
38.            <artifactId>kotlin-reflect</artifactId>
39.        </dependency>
40.        <dependency>
41.            <groupId>org.jetbrains.kotlin</groupId>
42.            <artifactId>kotlin-stdlib-jdk8</artifactId>
43.        </dependency>
44.        <dependency>
45.            <groupId>org.jetbrains.kotlinx</groupId>
46.            <artifactId>kotlinx-coroutines-core</artifactId>
47.            <version>1.3.2</version>
48.        </dependency>
49.    </dependencies>
50.    <build>
51.        <sourceDirectory>${project.basedir}/src/main/kotlin</sourceDirectory>
52.        <testSourceDirectory>${project.basedir}/src/test/kotlin</testSourceDirectory>
53.        <plugins>
54.            <plugin>
55.                <groupId>org.springframework.boot</groupId>
56.                <artifactId>spring-boot-maven-plugin</artifactId>
57.            </plugin>
58.            <plugin>
59.                <groupId>org.jetbrains.kotlin</groupId>
60.                <artifactId>kotlin-maven-plugin</artifactId>
```

```xml
61.        <configuration>
62.            <args>
63.                <arg>-Xjsr305=strict</arg>
64.            </args>
65.            <compilerPlugins>
66.                <plugin>spring</plugin>
67.                <plugin>jpa</plugin>
68.            </compilerPlugins>
69.        </configuration>
70.        <dependencies>
71.            <dependency>
72.                <groupId>org.jetbrains.kotlin</groupId>
73.                <artifactId>kotlin-maven-allopen</artifactId>
74.                <version>${kotlin.version}</version>
75.            </dependency>
76.            <dependency>
77.                <groupId>org.jetbrains.kotlin</groupId>
78.                <artifactId>kotlin-maven-noarg</artifactId>
79.                <version>${kotlin.version}</version>
80.            </dependency>
81.        </dependencies>
82.      </plugin>
83.    </plugins>
84.  </build>
85. </project>
```

application.yml 的定义如下，使用 Nacos 做服务注册中心：

```yaml
1. server:
2.   port: 8103    #应用端口号
3. spring:
4.   application:
5.     name: skywalking-service2   #应用名
6.   cloud:
7.     nacos:
8.       discovery:
9.         server-addr: 127.0.0.1:8848   #Nacos 服务注册中心地址
```

SkyWalkingApp1.kt 启动一个 Spring Boot 应用：

```kotlin
// 开启服务发现注解
@SpringBootApplication
@EnableDiscoveryClient
class SkyWalkingApp1 {
    @Bean
    @LoadBalanced
    fun restTemplate(): RestTemplate {
        return RestTemplate()
    }
}
// 启动函数
fun main(args: Array<String>) {
    runApplication<SkyWalkingApp1>(*args)
}
```

SkyWalkingController.kt 提供一个服务接口"service2"：

```kotlin
@RestController
class SkyWalkingController {
    // 测试接口
    @GetMapping("service2")
    fun service2(): String {
        return "Hello, I'm SkyWalking"
    }
}
```

新建一个 Maven 子工程：chapter10-skywalking，这是一个服务消费者，pom 文件如下：

```xml
<?xml version="1.0" encoding="UTF-8"?>
<project xmlns="http://maven.apache.org/POM/4.0.0"
    xmlns:xsi="http://www.w3.org/2001/XMLSchema-instance"
    xsi:schemaLocation="http://maven.apache.org/POM/4.0.0
http://maven.apache.org/xsd/maven-4.0.0.xsd">
    <parent>
        <artifactId>kotlinspringboot</artifactId>
```

```xml
7.        <groupId>io.kang.kotlinspringboot</groupId>
8.        <version>0.0.1-SNAPSHOT</version>
9.    </parent>
10.    <modelVersion>4.0.0</modelVersion>
11.    <!-- 子工程名-->
12.    <artifactId>chapter10-skywalking</artifactId>
13.    <dependencies>
14.        <!-- Spring Boot Web 依赖包-->
15.        <dependency>
16.            <groupId>org.springframework.boot</groupId>
17.            <artifactId>spring-boot-starter-web</artifactId>
18.            <version>2.2.1.RELEASE</version>
19.        </dependency>
20.        <!-- Spring Cloud Nacos 依赖包-->
21.        <dependency>
22.            <groupId>com.alibaba.cloud</groupId>
23.            <artifactId>spring-cloud-starter-alibaba-nacos-discovery</artifactId>
24.            <version>2.1.1.RELEASE</version>
25.        </dependency>
26.        <!-- Spring Cloud Feign 依赖包-->
27.        <dependency>
28.            <groupId>org.springframework.cloud</groupId>
29.            <artifactId>spring-cloud-starter-openfeign</artifactId>
30.            <version>2.2.1.RELEASE</version>
31.        </dependency>
32.        <dependency>
33.            <groupId>com.fasterxml.jackson.module</groupId>
34.            <artifactId>jackson-module-kotlin</artifactId>
35.        </dependency>
36.        <dependency>
37.            <groupId>org.jetbrains.kotlin</groupId>
38.            <artifactId>kotlin-reflect</artifactId>
39.        </dependency>
40.        <dependency>
41.            <groupId>org.jetbrains.kotlin</groupId>
42.            <artifactId>kotlin-stdlib-jdk8</artifactId>
```

```xml
43.        </dependency>
44.        <dependency>
45.            <groupId>org.jetbrains.kotlinx</groupId>
46.            <artifactId>kotlinx-coroutines-core</artifactId>
47.            <version>1.3.2</version>
48.        </dependency>
49.    </dependencies>
50.    <build>
51.        <sourceDirectory>${project.basedir}/src/main/kotlin</sourceDirectory>
52.        <testSourceDirectory>${project.basedir}/src/test/kotlin</testSourceDirectory>
53.        <plugins>
54.            <plugin>
55.                <groupId>org.springframework.boot</groupId>
56.                <artifactId>spring-boot-maven-plugin</artifactId>
57.            </plugin>
58.            <plugin>
59.                <groupId>org.jetbrains.kotlin</groupId>
60.                <artifactId>kotlin-maven-plugin</artifactId>
61.                <configuration>
62.                    <args>
63.                        <arg>-Xjsr305=strict</arg>
64.                    </args>
65.                    <compilerPlugins>
66.                        <plugin>spring</plugin>
67.                        <plugin>jpa</plugin>
68.                    </compilerPlugins>
69.                </configuration>
70.                <dependencies>
71.                    <dependency>
72.                        <groupId>org.jetbrains.kotlin</groupId>
73.                        <artifactId>kotlin-maven-allopen</artifactId>
74.                        <version>${kotlin.version}</version>
75.                    </dependency>
76.                    <dependency>
77.                        <groupId>org.jetbrains.kotlin</groupId>
78.                        <artifactId>kotlin-maven-noarg</artifactId>
```

```xml
79.                    <version>${kotlin.version}</version>
80.                </dependency>
81.            </dependencies>
82.        </plugin>
83.    </plugins>
84. </build>
85. </project>
```

application.yml 文件中的定义如下，使用 Nacos 做服务发现：

```yaml
1. server:
2.   port: 8102   # 应用端口号
3. spring:
4.   application:
5.     name: skywalking-service1   # 应用名
6.   cloud:
7.     nacos:
8.       discovery:
9.         server-addr: 127.0.0.1:8848   # Nacos 服务注册中心地址
```

SkyWalkingApp.kt 启动一个 Spring Boot 服务：

```kotlin
1. // 开启服务注册注解
2. @SpringBootApplication
3. @EnableDiscoveryClient
4. class SkyWalkingApp {
5.     @Bean
6.     @LoadBalanced
7.     fun restTemplate(): RestTemplate {
8.         return RestTemplate()
9.     }
10. }
11. // 启动函数
12. fun main(args: Array<String>) {
13.     runApplication<SkyWalkingApp>(*args)
14. }
```

SkyWalkingController.kt 调用"service2"接口：

```kotlin
1.  @RestController
2.  class SkyWalkingController {
3.      @Autowired
4.      lateinit var restTemplate: RestTemplate
5.      // 测试接口，通过 Ribbon 方式调用 service2 接口
6.      @GetMapping("service1")
7.      fun service1(): String {
8.          return restTemplate.getForObject("http://skywalking-service2/service2",
    String::class)
9.      }
10. }
```

分别在 VM Options 填写如下参数，然后运行 SkyWalkingApp1.kt、SkyWalkingApp.kt，启动服务提供方和服务消费方，通过代理方式，对代码无侵入，可监控服务间的调用。调用"service1"接口几次：

```
1.  -javaagent:D:\soft\apache-skywalking-apm-6.6.0\apache-skywalking-apm-bin\
    agent\skywalking-agent.jar
2.  -Dskywalking.agent.service_name=skywalking-service1
3.  -Dskywalking.collector.backend_service=localhost:11800
4.
5.  -javaagent:D:\soft\apache-skywalking-apm-6.6.0\apache-skywalking-apm-bin\
    agent\skywalking-agent.jar
6.  -Dskywalking.agent.service_name=skywalking-service2
7.  -Dskywalking.collector.backend_service=localhost:11800
```

在 SkyWalking 监控界面可以看到有两个 service，分别是 skywalking-service1 和 skywalking-service2，和 VM Options 参数定义的一致，如图 9.11 所示。

第 9 章　Kotlin 集成服务监控和服务链路监控

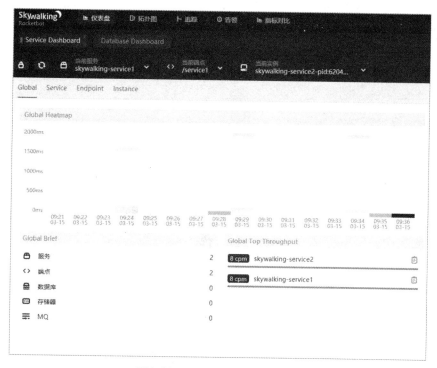

图9.11　SkyWalking监控界面

可以在图 9.12 中看到每次调用经过哪些接口和相应的耗时。

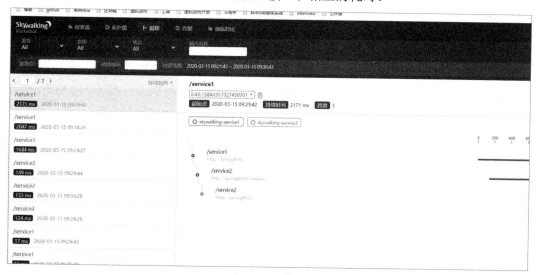

图9.12　SkyWalking监控接口调用耗时

在图 9.13 中可以看到服务的拓扑关系。

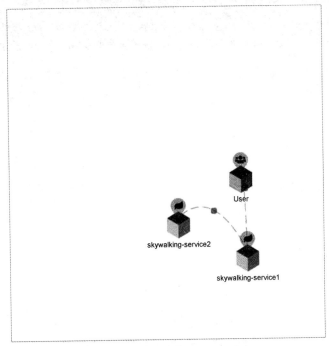

图9.13 服务拓扑关系

9.5 小结

本章介绍了使用 Kotlin 集成 Prometheus 和 Grafana 进行服务监控的方法。Prometheus 可以定时采集服务性能指标，Grafana 提供监控模板和优美的监控界面。本书使用 Micrometer 模板，提供 JVM 各种监控指标，此外还定义了一个接口调用次数监控指标。

本章还介绍了服务监控工具 ZipKin 和 Skywalking，它们有助于监控系统的调用情况，找到系统性能瓶颈，并有针对性地提高系统性能。

第 10 章
基于 Kotlin 和 Spring Boot 搭建博客

本章将在之前章节的基础上,通过一个完整的项目,介绍如何使用 Kotlin 和 Spring Boot 开发微服务系统。本章以一个博客系统为例,介绍初始化 Maven 工程、系统整体架构、数据实体定义、数据库设计、数据库操作层、Service 层及 Controller 层的相关知识。最后,介绍如何把该项目部署到腾讯云,并展示博客的效果。通过这个项目,读者可以搭建属于自己的博客平台,记录生活中的点滴。

10.1 初始化 Maven 工程

基于 Kotlin 和 Spring Boot 搭建博客,会用到表 10.1 列出的各项技术。

表 10.1 博客使用的技术栈列表

技术栈	介绍
Thymeleaf	模板引擎
SpringBoot	Spring 快速开发框架
Spring Session	session 管理
Kaptcha	验证码
CommonMark	Markdown 工具
MySQL	数据库

续表

技术栈	介绍
Spring Data Jpa	持久化层框架
QueryDSL	通用查询框架
Kotlin	开发语言

新建 Maven 子工程 chapter11-blog，pom.xml 如下：

```xml
1.  <?xml version="1.0" encoding="UTF-8"?>
2.  <project xmlns="http://maven.apache.org/POM/4.0.0"
3.          xmlns:xsi="http://www.w3.org/2001/XMLSchema-instance"
4.          xsi:schemaLocation="http://maven.apache.org/POM/4.0.0 http://maven.apache.org/xsd/maven-4.0.0.xsd">
5.      <parent>
6.          <artifactId>kotlinspringboot</artifactId>
7.          <groupId>io.kang.kotlinspringboot</groupId>
8.          <version>0.0.1-SNAPSHOT</version>
9.      </parent>
10.     <modelVersion>4.0.0</modelVersion>
11.     <!-- 子工程名-->
12.     <artifactId>chapter11-blog</artifactId>
13.     <dependencies>
14.         <!-- Spring Boot Web 依赖包-->
15.         <dependency>
16.             <groupId>org.springframework.boot</groupId>
17.             <artifactId>spring-boot-starter-web</artifactId>
18.         </dependency>
19.         <!-- Spring Boot Thymeleaf 模板引擎依赖包-->
20.         <dependency>
21.             <groupId>org.springframework.boot</groupId>
22.             <artifactId>spring-boot-starter-thymeleaf</artifactId>
23.         </dependency>
24.         <!-- Spring Session 依赖包-->
25.         <dependency>
26.             <groupId>org.springframework.session</groupId>
27.             <artifactId>spring-session-core</artifactId>
28.         </dependency>
```

```xml
29.    <!-- 验证码 -->
30.    <dependency>
31.        <groupId>com.github.penggle</groupId>
32.        <artifactId>kaptcha</artifactId>
33.        <version>2.3.2</version>
34.    </dependency>
35.    <!-- CommonMark Core 依赖包-->
36.    <dependency>
37.        <groupId>com.atlassian.commonmark</groupId>
38.        <artifactId>commonmark</artifactId>
39.        <version>0.8.0</version>
40.    </dependency>
41.    <!-- CommonMark Table 依赖包-->
42.    <dependency>
43.        <groupId>com.atlassian.commonmark</groupId>
44.        <artifactId>commonmark-ext-gfm-tables</artifactId>
45.        <version>0.8.0</version>
46.    </dependency>
47.    <!-- MySQL 驱动依赖包-->
48.    <dependency>
49.        <groupId>mysql</groupId>
50.        <artifactId>mysql-connector-java</artifactId>
51.        <scope>runtime</scope>
52.    </dependency>
53.    <!-- Spring Data JPA 依赖包-->
54.    <dependency>
55.        <groupId>org.springframework.boot</groupId>
56.        <artifactId>spring-boot-starter-data-jpa</artifactId>
57.    </dependency>
58.    <!-- QueryDSL 依赖包-->
59.    <dependency>
60.        <groupId>com.querydsl</groupId>
61.        <artifactId>querydsl-jpa</artifactId>
62.    </dependency>
63.    <dependency>
64.        <groupId>com.querydsl</groupId>
65.        <artifactId>querydsl-apt</artifactId>
```

```xml
66.        </dependency>
67.        <!-- Spring Boot Test 依赖包-->
68.        <dependency>
69.            <groupId>org.springframework.boot</groupId>
70.            <artifactId>spring-boot-starter-test</artifactId>
71.            <scope>test</scope>
72.        </dependency>
73.        <dependency>
74.            <groupId>com.fasterxml.jackson.module</groupId>
75.            <artifactId>jackson-module-kotlin</artifactId>
76.        </dependency>
77.        <dependency>
78.            <groupId>org.jetbrains.kotlin</groupId>
79.            <artifactId>kotlin-reflect</artifactId>
80.        </dependency>
81.        <dependency>
82.            <groupId>org.jetbrains.kotlin</groupId>
83.            <artifactId>kotlin-stdlib-jdk8</artifactId>
84.        </dependency>
85.        <dependency>
86.            <groupId>org.jetbrains.kotlinx</groupId>
87.            <artifactId>kotlinx-coroutines-core</artifactId>
88.            <version>1.3.2</version>
89.        </dependency>
90.    </dependencies>
91.    <build>
92.        <sourceDirectory>${project.basedir}/src/main/kotlin</sourceDirectory>
93.        <testSourceDirectory>${project.basedir}/src/test/kotlin</testSourceDirectory>
94.        <plugins>
95.            <plugin>
96.                <groupId>org.springframework.boot</groupId>
97.                <artifactId>spring-boot-maven-plugin</artifactId>
98.            </plugin>
99.            <plugin>
100.                <groupId>com.querydsl</groupId>
101.                <artifactId>querydsl-maven-plugin</artifactId>
102.                <executions>
```

```xml
            <execution>
                <phase>compile</phase>
                <goals>
                    <goal>jpa-export</goal>
                </goals>
                <configuration>
                    <targetFolder>target/generated-sources/kotlin</targetFolder>
                    <packages>io.kang.blog.entity</packages>
                </configuration>
            </execution>
        </executions>
    </plugin>
    <plugin>
        <groupId>org.jetbrains.kotlin</groupId>
        <artifactId>kotlin-maven-plugin</artifactId>
        <configuration>
            <args>
                <arg>-Xjsr305=strict</arg>
            </args>
            <compilerPlugins>
                <plugin>spring</plugin>
                <plugin>jpa</plugin>
            </compilerPlugins>
        </configuration>
        <dependencies>
            <dependency>
                <groupId>org.jetbrains.kotlin</groupId>
                <artifactId>kotlin-maven-allopen</artifactId>
                <version>${kotlin.version}</version>
            </dependency>
            <dependency>
                <groupId>org.jetbrains.kotlin</groupId>
                <artifactId>kotlin-maven-noarg</artifactId>
                <version>${kotlin.version}</version>
            </dependency>
        </dependencies>
    </plugin>
```

```
140.        </plugins>
141.    </build>
142.</project>
```

application.yml 文件的内容如下,其中定义了数据库连接信息、数据库连接池配置及应用名称:

```
1.  server:
2.    port: 8111   #应用端口号
3.  spring:
4.    thymeleaf:
5.      cache: false  #Thymeleaf 不开启缓存
6.    datasource:
7.      driver-class-name: com.mysql.cj.jdbc.Driver  #数据库驱动
8.      url: jdbc:mysql://localhost:3306/my_blog_db?useUnicode=
    true&characterEncoding=utf8&autoReconnect=true&useSSL=false&serverTimezone=
    UTC   #数据库连接串
9.      username: root     #数据库连接用户名
10.     password: 123456   #数据库连接密码
11.     hikari:
12.       minimum-idle: 5    #连接池空闲连接的最小数量
13.       maximum-pool-size: 15   #最大连接池大小
14.       auto-commit: true   #自动提交
15.       idle-timeout: 30000  # 空闲时间,30s
16.       pool-name: hikariCP   #连接池名称
17.       max-lifetime: 1800000  # 一个连接到连接池的存活时间,30 分钟
18.       connection-timeout: 30000  # 数据库连接超时时间,30s
19.   application:
20.     name: Blog   #应用名称
```

10.2 系统架构

本章介绍的博客的系统架构如图 10.1 所示。自下而上分为基础资源层、实体层、数据操作层、服务层和接口层。博客依赖 MySQL 数据库。

实体层分为管理用户实体、博客实体、博客分类实体、博客评论实体、博客配置实体、博客链接实体和博客标签实体,每个实体对应数据库中的一张表,每个实体描述博客系统的不同维度属性。

每个实体有对应的数据操作 DAO，本书使用 JPA 和 QueryDSL 实现管理用户、博客、博客分类、博客评论、博客配置、博客链接及博客标签的查询、更新、插入等基础操作和分页查询、批量更新等高级操作。

服务层在实体层和数据操作层的基础上提供相关的操作。根据功能，服务层划分为管理用户服务、博客服务、博客分类服务、博客评论服务、博客配置服务、博客链接服务及博客标签服务。

接口层提供前台展示接口和后台管理接口。前台展示接口主要提供首页博客展示、类别展示、详情展示、标签列表展示、评论操作、友情链接展示、关于页面展示等功能。后台管理接口提供管理用户相关、博客编辑相关、发布相关、类别管理相关、评论审核相关、配置相关、友情链接相关、标签相关及上传相关接口。

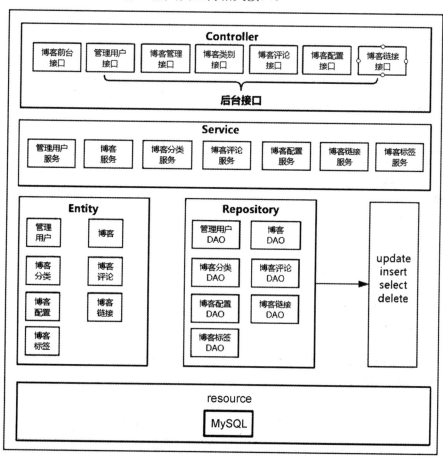

图10.1　博客系统架构图

10.3 定义实体

博客系统定义了表 10.2 所示的实体。

表 10.2　博客定义的实体列表

序号	实体	说明	属性
1	AdminUser	管理用户	用户 id、登录用户名、登录密码、昵称、是否锁定
2	Blog	博客	博客 id、标题、url、封面图片、分类 id、分类名称、标签 id、博客审核状态、点击量、是否允许评论、是否删除、创建时间、更新时间、博客内容
3	BlogCategory	博客分类	分类 id、分类名称、图标、排名、是否删除、创建时间
4	BlogComment	博客评论	评论 id、博客 id、评论者、邮箱、网址、评论内容、创建时间、评论者 IP 地址、回复内容、回复时间、评论审核状态、是否删除
5	BlogConfig	博客配置	配置名称、配置值、创建时间、更新时间
6	BlogLink	博客链接	友链 id、类型、名称、url、描述、排名、是否删除、创建时间
7	BlogTag	博客标签	标签 id、名称、是否删除、创建时间
8	BlogTagRelation	博客标签关系	关系 id、博客 id、标签 id、创建时间

AdminUser.kt 定义了管理用户实体，包括 adminUserId、loginUserName、loginPassword、nickName、locked 这几个属性。对应数据库中的 tb_admin_user 表，代码如下：

```
1.  // 实体，对应数据库中的 tb_admin_user 表
2.  @Entity
3.  @Table(name = "tb_admin_user")
4.  class AdminUser {
5.      // 用户 id
6.      @Id
7.      @GeneratedValue(strategy = GenerationType.IDENTITY)
8.      var adminUserId: Int? = null
9.      // 登录用户名
10.     var loginUserName: String = ""
11.         set(loginUserName) {
12.             field = loginUserName?.trim { it <= ' ' }
13.         }
```

```kotlin
14.    // 登录密码
15.    var loginPassword: String = ""
16.        set(loginPassword) {
17.            field = loginPassword?.trim { it <= ' ' }
18.        }
19.    // 昵称
20.    var nickName: String = ""
21.        set(nickName) {
22.            field = nickName?.trim { it <= ' ' }
23.        }
24.    // 是否锁定
25.    var locked: Byte = 0
26. }
```

Blog.kt 定义了博客（文章）实体，包括 blogId、blogTitle、blogSubUrl、blogCoverImage、blogCategoryId、blogCategoryName、blogTags、blogStatus、blogViews、enableComment、isDeleted、createTime、updateTime、blogContent 这几个属性，对应数据库中的 tb_blog 表。代码如下：

```kotlin
1.  // blog 实体，对应数据库中的 tb_blog 表
2.  @Entity
3.  @Table(name = "tb_blog")
4.  class Blog {
5.      // 博客 id
6.      @Id
7.      @GeneratedValue(strategy = GenerationType.IDENTITY)
8.      var blogId: Long = 0
9.      // 博客标题
10.     var blogTitle: String = ""
11.         set(blogTitle) {
12.             field = blogTitle.trim { it <= ' ' }
13.         }
14.     // 博客 url
15.     var blogSubUrl: String = ""
16.         set(blogSubUrl) {
17.             field = blogSubUrl.trim { it <= ' ' }
18.         }
```

```kotlin
19.     // 博客封面图片
20.     var blogCoverImage: String = ""
21.         set(blogCoverImage) {
22.             field = blogCoverImage.trim { it <= ' ' }
23.         }
24.     // 博客分类 id
25.     var blogCategoryId: Int = 0
26.     // 博客分类名称
27.     var blogCategoryName: String = ""
28.         set(blogCategoryName) {
29.             field = blogCategoryName.trim { it <= ' ' }
30.         }
31.     // 博客标签
32.     var blogTags: String = ""
33.         set(blogTags) {
34.             field = blogTags.trim { it <= ' ' }
35.         }
36.     // 博客审核状态
37.     var blogStatus: Byte = 0
38.     // 博客点击量
39.     var blogViews: Long = 0
40.     // 博客是否允许评论
41.     var enableComment: Byte = 0
42.     // 博客是否被删除
43.     var isDeleted: Byte = 0
44.     // 博客创建时间
45.     @JsonFormat(pattern = "yyyy-MM-dd HH:mm:ss", timezone = "GMT+8")
46.     var createTime: Date = Date()
47.     // 博客更新时间
48.     var updateTime: Date = Date()
49.     // 博客内容
50.     var blogContent: String = ""
51.         set(blogContent) {
52.             field = blogContent.trim { it <= ' ' }
53.         }
54. }
```

BlogCategory.kt 定义了博客分类实体，包括 categoryId、categoryName、categoryIcon、 categoryRank、isDeleted、createTime 属性，对应数据库中的 tb_blog_category 表。代码如下：

```kotlin
// 博客分类实体，对应数据库中的 tb_blog_category 表
@Entity
@Table(name = "tb_blog_category")
class BlogCategory {
    // 博客分类 id
    @Id
    @GeneratedValue(strategy = GenerationType.IDENTITY)
    var categoryId: Int = 0
    // 博客分类名称
    var categoryName: String = ""
        set(categoryName) {
            field = categoryName.trim { it <= ' ' }
        }
    // 博客分类图标
    var categoryIcon: String = ""
        set(categoryIcon) {
            field = categoryIcon.trim { it <= ' ' }
        }
    // 博客分类排名
    var categoryRank: Int = 0
    // 博客分类是否被删除
    var isDeleted: Byte = 0
    @JsonFormat(pattern = "yyyy-MM-dd HH:mm:ss", timezone = "GMT+8")
    // 创建时间
    var createTime: Date = Date()
}
```

BlogComment.kt 定义了博客评论实体，包括 commentId、blogId、commentator、email、websiteUrl、commentBody、commentCreateTime、commentatorIp、replyBody、replyCreateTime、commentStatus、isDeleted 属性，对应数据库中的 tb_blog_comment 表。代码如下：

```kotlin
// 博客评论实体，对应数据库表中的 tb_blog_comment 表
@Entity
@Table(name = "tb_blog_comment")
class BlogComment {
```

```kotlin
5.      // 评论 id
6.      @Id
7.      @GeneratedValue(strategy = GenerationType.IDENTITY)
8.      var commentId: Long? = null
9.      // 博客 id
10.     var blogId: Long = 0
11.     // 评论人
12.     var commentator: String = ""
13.         set(commentator) {
14.             field = commentator.trim { it <= ' ' }
15.         }
16.     // 邮箱
17.     var email: String = ""
18.         set(email) {
19.             field = email.trim { it <= ' ' }
20.         }
21.     // 网址
22.     var websiteUrl: String = ""
23.         set(websiteUrl) {
24.             field = websiteUrl.trim { it <= ' ' }
25.         }
26.     // 评论内容
27.     var commentBody: String = ""
28.         set(commentBody) {
29.             field = commentBody.trim { it <= ' ' }
30.         }
31.     // 评论创建时间
32.     @JsonFormat(pattern = "yyyy-MM-dd HH:mm:ss", timezone = "GMT+8")
33.     var commentCreateTime: Date = Date()
34.     // 评论者 IP 地址
35.     var commentatorIp: String = ""
36.         set(commentatorIp) {
37.             field = commentatorIp.trim { it <= ' ' }
38.         }
39.     // 回复内容
40.     var replyBody: String = ""
41.         set(replyBody) {
```

```
42.         field = replyBody.trim { it <= ' ' }
43.     }
44.     // 回复创建时间
45.     @JsonFormat(pattern = "yyyy-MM-dd HH:mm:ss", timezone = "GMT+8")
46.     var replyCreateTime: Date = Date()
47.     // 评论状态
48.     var commentStatus: Byte = 0
49.     // 是否删除
50.     var isDeleted: Byte = 0
51. }
```

BlogConfig.kt 定义了博客配置实体，包括 configName、configValue、createTime、updateTime 属性，对应数据库中的 tb_config 表。代码如下：

```
1.  // 博客配置实体，对应数据库中的 tb_config 表
2.  @Entity
3.  @Table(name = "tb_config")
4.  class BlogConfig {
5.      // 配置名称
6.      @Id
7.      var configName: String = ""
8.          set(configName) {
9.              field = configName.trim { it <= ' ' }
10.         }
11.     // 配置值
12.     var configValue: String = ""
13.         set(configValue) {
14.             field = configValue.trim { it <= ' ' }
15.         }
16.     // 配置创建时间
17.     @JsonFormat(pattern = "yyyy-MM-dd HH:mm:ss", timezone = "GMT+8")
18.     var createTime: Date = Date()
19.     // 配置更新时间
20.     @JsonFormat(pattern = "yyyy-MM-dd HH:mm:ss", timezone = "GMT+8")
21.     var updateTime: Date = Date()
22. }
```

BlogLink.kt 定义了博客链接（友链）实体，包括 linkId、linkType、linkName、linkUrl、 linkDescription、linkRank、isDeleted、createTime 属性，对应数据库中的 tb_link 表。代码如下：

```kotlin
1.  // 友链实体，对应数据库中的 tb_link 表
2.  @Entity
3.  @Table(name = "tb_link")
4.  class BlogLink {
5.      // 友链 id
6.      @Id
7.      @GeneratedValue(strategy = GenerationType.IDENTITY)
8.      var linkId: Int? = null
9.      // 友链类型
10.     var linkType: Byte = 0
11.     // 友链名称
12.     var linkName: String = ""
13.         set(linkName) {
14.             field = linkName.trim { it <= ' ' }
15.         }
16.     // 友链地址
17.     var linkUrl: String = ""
18.         set(linkUrl) {
19.             field = linkUrl.trim { it <= ' ' }
20.         }
21.     // 友链描述
22.     var linkDescription: String = ""
23.         set(linkDescription) {
24.             field = linkDescription.trim { it <= ' ' }
25.         }
26.     // 友链排名
27.     var linkRank: Int = 0
28.     // 友链是否删除
29.     var isDeleted: Byte = 0
30.     // 友链创建时间
31.     @JsonFormat(pattern = "yyyy-MM-dd HH:mm:ss", timezone = "GMT+8")
32.     var createTime: Date = Date()
33. }
```

BlogTag.kt 定义了博客标签实体，包括 tagId、tagName、isDeleted、createTime 属性，

对应数据库中的 tb_blog_tag 表。代码如下：

```kotlin
// 标签实体，对应数据库中的 tb_blog_tag 表
@Entity
@Table(name = "tb_blog_tag")
class BlogTag {
    // 标签 id
    @Id
    @GeneratedValue(strategy = GenerationType.IDENTITY)
    var tagId: Int = 0
    // 标签名称
    var tagName: String = ""
        set(tagName) {
            field = tagName.trim { it <= ' ' }
        }
    // 标签是否删除
    var isDeleted: Byte = 0
    // 标签创建时间
    @JsonFormat(pattern = "yyyy-MM-dd HH:mm:ss", timezone = "GMT+8")
    var createTime: Date = Date()
}
```

BlogTagRelation.kt 定义了博客标签关系实体，包括 relationId、blogId、tagId、createTime 属性，对应数据库中的 tb_blog_tag_relation 表。代码如下：

```kotlin
// 博客标签关系实体，对应数据库 tb_blog_tag_relation 表
@Entity
@Table(name = "tb_blog_tag_relation")
class BlogTagRelation {
    // 博客标签关系 id
    @Id
    @GeneratedValue(strategy = GenerationType.IDENTITY)
    var relationId: Long? = null
    // 博客 id
    var blogId: Long = 0
    // 标签 id
    var tagId: Int = 0
    // 创建时间
    @JsonFormat(pattern = "yyyy-MM-dd HH:mm:ss", timezone = "GMT+8")
```

```
15.     var createTime: Date = Date()
16. }
```

10.4 数据库设计

10.3 节中的实体对应不同的数据库表：tb_admin_user、tb_blog、tb_blog_category、tb_blog_comment、tb_config、tb_link、tb_blog_tag、tb_blog_tag_relation。数据库表的关系如图 10.2 所示。

图10.2 数据库关系图

10.5 Repository 层的设计

博客系统定义的 Repository/DAO 如表 10.3 所示。

表 10.3 博客定义的 Repository/DAO 列表

序号	Repository/DAO	说明	方法
1	AdminUserRepository	管理用户 JPA 接口	提供增删改查方法
2	BlogCategoryRepository	分类 JPA 接口	提供增删改查方法
3	BlogCommentRepository	评论 JPA 接口	提供增删改查方法
4	BlogConfigRepository	配置 JPA 接口	提供增删改查方法
5	BlogLinkRepository	链接 JPA 接口	提供增删改查方法
6	BlogRepository	博客 JPA 接口	提供增删改查方法
7	BlogTagRelationRepository	博客标签关系 JPA 接口	提供增删改查方法
8	BlogTagRepository	标签 JPA 接口	提供增删改查方法
9	AdminUserDAO	用户 DAO 类	新增用户、登录、根据 userId 查询、更新记录
10	BlogCategoryDAO	分类 DAO 类	逻辑删除、新增分类、查询分类、更新字段、分页查询、批量查询、批量删除、查询总数
11	BlogCommentDAO	评论 DAO 类	逻辑删除、新增评论、查询评论、选择性更新、分页查找、查询总数、批量删除
12	BlogConfigDAO	配置 DAO 类	查询所有配置、根据配置名称查询、选择性更新
13	BlogDAO	博客 DAO 类	删除、保存博客、根据博客 id 查询博客、选择性更新博客、分页查询博客、根据类型查询博客、查询总数、批量删除、根据标签分页查询、更新分类
14	BlogLinkDAO	链接 DAO 类	删除友链、新增友链、查找友链、选择性更新友链、分页查找友链、查询总数、批量删除
15	BlogTagDAO	标签 DAO 类	删除标签、新增标签、查询标签、选择性更新标签、分页查询标签、查询总数、批量删除
16	BlogTagRelationDAO	博客标签关系 DAO 类	删除、保存、查询、选择性更新、批量插入、批量删除

AdminUserRepository.kt、BlogCategoryRepository.kt、BlogCommentRepository.kt、BlogConfigRepository.kt、BlogLinkRepository.kt、BlogRepository.kt、BlogTagRelationRepository.kt、BlogTagRepository.kt 都继承了 JpaRepository，JPA 提供了增删改查方法。但是 JPA 对原生 SQL 支持不太好，QueryDSL 提供了便利。

```kotlin
// 操作 AdminUser 对应的数据库表的 JpaRepository 接口
interface AdminUserRepository : JpaRepository<AdminUser, Long>,
    QuerydslPredicateExecutor<AdminUser>
// 操作 BlogCategory 对应的数据库表的 JpaRepository 接口
interface BlogCategoryRepository : JpaRepository<BlogCategory, Long>,
    QuerydslPredicateExecutor<BlogCategory>
// 操作 BlogComment 对应的数据库表的 JpaRepository 接口
interface BlogCommentRepository : JpaRepository<BlogComment, Long>,
    QuerydslPredicateExecutor<BlogComment>
// 操作 BlogConfig 对应的数据库表的 JpaRepository 接口
interface BlogConfigRepository : JpaRepository<BlogConfig, String>,
    QuerydslPredicateExecutor<BlogConfig>
// 操作 BlogLink 对应的数据库表的 JpaRepository 接口
interface BlogLinkRepository : JpaRepository<BlogLink, Long>,
    QuerydslPredicateExecutor<BlogLink>
// 操作 Blog 对应的数据库表的 JpaRepository 接口
interface BlogRepository : JpaRepository<Blog, Long>,
    QuerydslPredicateExecutor<Blog>
// 操作 BlogTagRelation 对应的数据库表的 JpaRepository 接口
interface BlogTagRelationRepository : JpaRepository<BlogTagRelation, Long>,
    QuerydslPredicateExecutor<BlogTagRelation>
// 操作 BlogTag 对应的数据库表的 JpaRepository 接口
interface BlogTagRepository : JpaRepository<BlogTag, Long>,
    QuerydslPredicateExecutor<BlogTag>
```

AdminUserDAO.kt 基于 QueryDSL 实现了新增用户；根据用户名、密码校验该用户是否存在，模拟登录行为；根据用户 id 查询用户；对字段进行选择性更新，当字段不为 null 才会更新相关记录的字段；根据用户 id，更新指定用户的字段。代码如下：

```kotlin
@Component
class AdminUserDAO {
    @Autowired
```

```kotlin
4.     lateinit var adminUserRepository: AdminUserRepository
5.     @Autowired
6.     lateinit var queryFactory: JPAQueryFactory
7.     @Transactional
8.     // 保存一条 AdminUser 数据
9.     fun insert(record: AdminUser): Int {
10.        adminUserRepository.save(record)
11.        return 0
12.    }
13.    // 保存一条 AdminUser 数据
14.    @Transactional
15.    fun insertSelective(record: AdminUser): Int {
16.        adminUserRepository.save(record)
17.        return 0
18.    }
19.    // 根据 userName、password 查询是否存在一条 AdminUser
20.    fun login(userName: String, password: String): AdminUser? {
21.        val qAdminUser = QAdminUser.adminUser
22.        return queryFactory.selectFrom(qAdminUser)
23.                .where(qAdminUser.loginUserName.eq(userName).and(qAdminUser.loginPassword.eq(password)))
24.                .fetchOne()
25.    }
26.    // 根据 adminUserId 查找 AdminUser
27.    fun selectByPrimaryKey(adminUserId: Int): AdminUser? {
28.        val qAdminUser = QAdminUser.adminUser
29.        return queryFactory.selectFrom(qAdminUser)
30.                .where(qAdminUser.adminUserId.eq(adminUserId))
31.                .fetchOne()
32.    }
33.    // 将 record 中不为 null 的字段更新到数据库
34.    @Transactional
35.    fun updateByPrimaryKeySelective(record: AdminUser): Int {
36.        val qAdminUser = QAdminUser.adminUser
37.        val cols = arrayListOf<Path<*>>()
38.        val values = arrayListOf<Any?>()
39.        if(record.loginUserName != null) {
```

```kotlin
40.            cols.add(qAdminUser.loginUserName)
41.            values.add(record.loginUserName)
42.        }
43.        if(record.loginPassword != null) {
44.            cols.add(qAdminUser.loginPassword)
45.            values.add(record.loginPassword)
46.        }
47.        if(record.nickName != null) {
48.            cols.add(qAdminUser.nickName)
49.            values.add(record.nickName)
50.        }
51.        if(record.locked != null) {
52.            cols.add(qAdminUser.locked)
53.            values.add(record.locked)
54.        }
55.        return queryFactory.update(qAdminUser)
56.                .set(cols, values)
57.                .where(qAdminUser.adminUserId.eq(record.adminUserId))
58.                .execute()
59.                .toInt()
60.    }
61.    // 更新 AdminUser 数据
62.    @Transactional
63.    fun updateByPrimaryKey(record: AdminUser): Int {
64.        val qAdminUser = QAdminUser.adminUser
65.        return queryFactory.update(qAdminUser)
66.                .set(qAdminUser.loginUserName, record.loginUserName)
67.                .set(qAdminUser.loginPassword, record.loginPassword)
68.                .set(qAdminUser.nickName, record.nickName)
69.                .set(qAdminUser.locked, record.locked)
70.                .where(qAdminUser.adminUserId.eq(record.adminUserId))
71.                .execute()
72.                .toInt()
73.    }
74. }
```

BlogCategoryDAO.kt 基于 QueryDSL 实现了根据分类 id 删除该条记录，把该记录的

isDeleted 更新为 1，表示删除；插入一条 BlogCategory；根据 categoryId 查询 BlogCategory；根据 categoryName 查询 BlogCategory；BlogCategory 属性不为 null，更新数据库相应字段；根据 categoryId 更新该分类相关字段；按照 categoryRank、createTime 降序分页查询有效的博客分类；根据 categoryId 批量查询分类；查询分类总数；将批量更新记录的 isDeleted 设为 1。代码如下：

```kotlin
1.  @Component
2.  class BlogCategoryDAO {
3.      @Autowired
4.      lateinit var queryFactory: JPAQueryFactory
5.      @Autowired
6.      lateinit var blogCategoryRepository: BlogCategoryRepository
7.      // 根据 categoryId 删除
8.      @Transactional
9.      fun deleteByPrimaryKey(categoryId: Int): Int {
10.         // 省略部分代码……
11.         return queryFactory.update(qBlogCategory)
12.                 .set(qBlogCategory.isDeleted, 1)
13.                 .where(predicate)
14.                 .execute()
15.                 .toInt()
16.     }
17.     // 保存一条 BlogCategory 数据
18.     @Transactional
19.     fun insert(record: BlogCategory): Int {
20.         blogCategoryRepository.save(record)
21.         return 0
22.     }
23.     // 保存一条 BlogCategory 数据
24.     @Transactional
25.     fun insertSelective(record: BlogCategory): Int {
26.         blogCategoryRepository.save(record)
27.         return 0
28.     }
29.     // 根据 categoryId 查找 BlogCategory
30.     fun selectByPrimaryKey(categoryId: Int?): BlogCategory? {
```

```kotlin
31.         // 省略部分代码……
32.         val predicate = qBlogCategory.categoryId.eq(categoryId)
33.             .and(qBlogCategory.isDeleted.eq(0))
34.     }
35.     // 根据 categoryName 查找 BlogCategory
36.     fun selectByCategoryName(categoryName: String): BlogCategory? {
37.         // 省略部分代码……
38.         val predicate = qBlogCategory.categoryName.eq(categoryName)
39.             .and(qBlogCategory.isDeleted.eq(0))
40.     }
41.     // 将 record 不为 null 的字段更新到数据库
42.     @Transactional
43.     fun updateByPrimaryKeySelective(record: BlogCategory): Int {
44.         // 省略部分代码……
45.         return queryFactory.update(qBlogCategory)
46.             .set(cols, values)
47.             .where(qBlogCategory.categoryId.eq(record.categoryId))
48.             .execute()
49.             .toInt()
50.     }
51.     // 更新 BlogCategory
52.     @Transactional
53.     fun updateByPrimaryKey(record: BlogCategory): Int {
54.         val qBlogCategory = QBlogCategory.blogCategory
55.         return queryFactory.update(qBlogCategory)
56.             .set(qBlogCategory.categoryName, record.categoryName)
57.             .set(qBlogCategory.categoryIcon, record.categoryIcon)
58.             .set(qBlogCategory.categoryRank, record.categoryRank)
59.             .set(qBlogCategory.isDeleted, record.isDeleted)
60.             .set(qBlogCategory.createTime, record.createTime)
61.             .where(qBlogCategory.categoryId.eq(record.categoryId))
62.             .execute()
63.             .toInt()
64.     }
65.     // 分页查找 BlogCategory
66.     fun findCategoryList(pageUtil: PageQueryUtil?): List<BlogCategory> {
67.         // 省略部分代码……
```

```kotlin
68.    }
69.    // 根据 categoryIds 查找 BlogCategory
70.    fun selectByCategoryIds(categoryIds: List<Int?>): List<BlogCategory> {
71.        // 省略部分代码……
72.        val predicate = qBlogCategory.categoryId.`in`(categoryIds)
73.            .and(qBlogCategory.isDeleted.eq(0))
74.    }
75.    // 查询总数
76.    fun getTotalCategories(pageUtil: PageQueryUtil?): Int {
77.        // 省略部分代码……
78.        return queryFactory.selectFrom(qBlogCategory)
79.            .where(qBlogCategory.isDeleted.eq(0))
80.            .fetchCount()
81.            .toInt()
82.    }
83.    // 批量删除 BlogCategory
84.    @Transactional
85.    fun deleteBatch(ids: List<Int>): Int {
86.        // 省略部分代码……
87.        val predicate = qBlogCategory.categoryId.`in`(ids)
88.        return queryFactory.update(qBlogCategory)
89.            .set(qBlogCategory.isDeleted, 1)
90.            .where(predicate)
91.            .execute()
92.            .toInt()
93.    }
94. }
```

BlogCommentDAO.kt 基于 QueryDSL 实现了根据 commentId 物理删除评论；插入一条评论；根据 commentId 查询该评论；当字段不为 null 时更新数据库对应字段；根据 commentId 更新对应的记录；按照 commentId 降序分页查找评论；根据 blogId、commentStatus 查询评论总数；将一组评论的 commentStatus 更新为 1；批量删除评论。代码如下：

```kotlin
1. @Component
2. class BlogCommentDAO {
3.     @Autowired
4.     lateinit var queryFactory: JPAQueryFactory
```

```kotlin
5.  @Autowired
6.  lateinit var blogCommentRepository: BlogCommentRepository
7.  // 根据 commentId 删除
8.  @Transactional
9.  fun deleteByPrimaryKey(commentId: Long): Int {
10.     // 省略部分代码……
11. }
12. // 保存一条 BlogComment
13. @Transactional
14. fun insert(record: BlogComment): Int {
15.     blogCommentRepository.save(record)
16.     return 0
17. }
18. // 保存一条 BlogComment
19. @Transactional
20. fun insertSelective(record: BlogComment): Int {
21.     blogCommentRepository.save(record)
22.     return 0
23. }
24. // 根据 commentId 查找 BlogComment
25. fun selectByPrimaryKey(commentId: Long): BlogComment? {
26.     val qBlogComment = QBlogComment.blogComment
27.     return queryFactory.selectFrom(qBlogComment)
28.         .where(qBlogComment.commentId.eq(commentId).and(qBlogComment.isDeleted.eq(0)))
29.         .fetchOne()
30. }
31. // 将 record 中不为 null 的字段更新到数据库
32. @Transactional
33. fun updateByPrimaryKeySelective(record: BlogComment): Int {
34.     // 省略部分代码……
35.     return queryFactory.update(qBlogComment)
36.         .set(cols, values)
37.         .where(qBlogComment.commentId.eq(record.commentId))
38.         .execute()
39.         .toInt()
40. }
```

```kotlin
41.    // 更新 BlogComment
42.    @Transactional
43.    fun updateByPrimaryKey(record: BlogComment): Int {
44.        // 省略部分代码……
45.    }
46.    // 分页查找 BlogComment
47.    fun findBlogCommentList(map: Map<*, *>): List<BlogComment> {
48.        // 省略部分代码……
49.        return if(start != null && limit != null) {
50.            queryFactory.selectFrom(qBlogComment)
51.                    .where(predicate)
52.                    .where(predicate1)
53.                    .where(qBlogComment.isDeleted.eq(0))
54.                    .orderBy(OrderSpecifier(Order.DESC, qBlogComment.commentId))
55.                    .offset(start.toLong())
56.                    .limit(limit.toLong())
57.                    .fetchResults()
58.                    .results
59.        }else {
60.            listOf()
61.        }
62.    }
63.    // 根据 blogId、CommentStatus 查询总数
64.    fun getTotalBlogComments(map: Map<*, *>?): Int {
65.        // 省略部分代码……
66.        return queryFactory.selectFrom(qBlogComment)
67.                .where(predicate)
68.                .where(predicate1)
69.                .where(qBlogComment.isDeleted.eq(0))
70.                .fetchCount()
71.                .toInt()
72.    }
73.    // 根据 ids，将 commentStatus 设置为 1
74.    @Transactional
75.    fun checkDone(ids: List<Long>): Int {
76.        val qBlogComment = QBlogComment.blogComment
77.        return queryFactory.update(qBlogComment)
```

```kotlin
78.                    .set(qBlogComment.commentStatus, 1)
79.                    .where(qBlogComment.commentId.`in`(ids).and
   (qBlogComment.commentStatus.eq(0)))
80.                    .execute()
81.                    .toInt()
82.        }
83.        // 根据ids，批量删除
84.        @Transactional
85.        fun deleteBatch(ids: List<Long>): Int {
86.            val qBlogComment = QBlogComment.blogComment
87.            return queryFactory.update(qBlogComment)
88.                    .set(qBlogComment.isDeleted, 1)
89.                    .where(qBlogComment.commentId.`in`(ids))
90.                    .execute()
91.                    .toInt()
92.        }
93.    }
```

BlogConfigDAO.kt 基于 QueryDSL 实现了查询所有博客配置；根据配置名称查询相应的配置；选择性更新不为 null 的配置属性等。代码如下：

```kotlin
1.    @Component
2.    class BlogConfigDAO {
3.        @Autowired
4.        lateinit var queryFactory: JPAQueryFactory
5.        // 查找所有数据
6.        fun selectAll(): List<BlogConfig> {
7.            val qBlogConfig = QBlogConfig.blogConfig
8.            return queryFactory.selectFrom(qBlogConfig)
9.                    .fetchResults()
10.                   .results
11.       }
12.       // 根据configName查找数据
13.       fun selectByPrimaryKey(configName: String): BlogConfig {
14.           val qBlogConfig = QBlogConfig.blogConfig
15.           val predicate = qBlogConfig.configName.eq(configName)
16.           return queryFactory.selectFrom(qBlogConfig)
```

```kotlin
17.             .where(predicate)
18.             .fetchFirst()
19.     }
20.     // 将 record 不为 null 的字段更新到数据库
21.     @Transactional
22.     fun updateByPrimaryKeySelective(record: BlogConfig): Int {
23.         // 省略部分代码……
24.         return queryFactory.update(qBlogConfig)
25.             .set(cols, values)
26.             .where(qBlogConfig.configName.eq(record.configName))
27.             .execute()
28.             .toInt()
29.     }
30. }
```

BlogDAO.kt 基于 QueryDSL 实现了删除指定 blogId 的博客；保存一条博客；根据 blogId 查询博客；选择性更新不为 null 的博客属性；根据 blogId 更新指定博客的属性；根据关键字、状态、分类按照 blogId 降序分页查找博客；根据 blogId 或者 blogViews 降序查询博客；根据关键字、状态、分类查询博客总数；批量删除博客；按照 blogId 降序分页查找指定标签的博客；查询指定标签的博客总数；根据 url 查询博客；更新博客的标签和分类。代码如下：

```kotlin
1. @Component
2. class BlogDAO {
3.     @Autowired
4.     lateinit var queryFactory: JPAQueryFactory
5.     @Autowired
6.     lateinit var blogRepository: BlogRepository
7.     // 根据 blogId 删除数据
8.     @Transactional
9.     fun deleteByPrimaryKey(blogId: Long): Int {
10.         val qBlog = QBlog.blog
11.         return queryFactory.update(qBlog)
12.             .set(qBlog.isDeleted, 1)
13.             .where(qBlog.isDeleted.eq(0).and(qBlog.blogId.eq(blogId)))
14.             .execute()
15.             .toInt()
16.     }
```

```kotlin
17.     // 保存一条 Blog
18.     @Transactional
19.     fun insert(record: Blog): Int {
20.         blogRepository.save(record)
21.         return 0
22.     }
23.     // 保存一条 Blog
24.     @Transactional
25.     fun insertSelective(record: Blog): Int {
26.         blogRepository.save(record)
27.         return 0
28.     }
29.     // 根据 blogId 查找 Blog
30.     fun selectByPrimaryKey(blogId: Long?): Blog {
31.         val qBlog = QBlog.blog
32.         return queryFactory.selectFrom(qBlog)
33.                 .where(qBlog.blogId.eq(blogId))
34.                 .fetchFirst()
35.     }
36.     // 将 record 不为 null 的字段更新到数据库
37.     @Transactional
38.     fun updateByPrimaryKeySelective(record: Blog): Int {
39.         // 省略部分代码……
40.         return queryFactory.update(qBlog)
41.                 .set(cols, values)
42.                 .where(qBlog.blogId.eq(record.blogId))
43.                 .execute()
44.                 .toInt()
45.     }
46.     // 更新 Blog
47.     @Transactional
48.     fun updateByPrimaryKeyWithBLOBs(record: Blog): Int {
49.         // 省略部分代码……
50.     }
51.     // 更新 Blog
52.     @Transactional
53.     fun updateByPrimaryKey(record: Blog): Int {
```

```kotlin
54.        // 省略部分代码……
55.    }
56.    // 分页查找 Blog
57.    fun findBlogList(pageUtil: PageQueryUtil): List<Blog> {
58.        // 省略部分代码……
59.        return queryFactory.selectFrom(qBlog)
60.                .where(predicate)
61.                .where(predicate1)
62.                .where(predicate2)
63.                .where(qBlog.isDeleted.eq(0))
64.                .orderBy(OrderSpecifier(Order.DESC, qBlog.blogId))
65.                .offset(start.toLong())
66.                .limit(limit.toLong())
67.                .fetchResults()
68.                .results
69.    }
70.    // 根据 blogId 或者 blogViews 降序查找 limit 条 Blog
71.    fun findBlogListByType(type: Int, limit: Int): List<Blog> {
72.        // 省略部分代码……
73.        return queryFactory.selectFrom(qBlog)
74.                .where(qBlog.isDeleted.eq(0).and(qBlog.blogStatus.eq(1)))
75.                .orderBy(order)
76.                .limit(limit.toLong())
77.                .fetchResults().results
78.    }
79.    // 查询满足条件的 blog 总数
80.    fun getTotalBlogs(pageUtil: PageQueryUtil?): Int {
81.        // 省略部分代码……
82.        return queryFactory.selectFrom(qBlog)
83.                .where(predicate)
84.                .where(predicate1)
85.                .where(predicate2)
86.                .where(qBlog.isDeleted.eq(0))
87.                .fetchCount().toInt()
88.    }
89.    // 根据 ids 批量删除 blog
90.    @Transactional
```

```kotlin
91.     fun deleteBatch(ids: List<Long>): Int {
92.         val qBlog = QBlog.blog
93.         return queryFactory.update(qBlog)
94.                 .set(qBlog.isDeleted, 1)
95.                 .where(qBlog.blogId.`in`(ids))
96.                 .execute()
97.                 .toInt()
98.     }
99.     // 根据 tagId，分页查找 Blog
100.    fun getBlogsPageByTagId(pageUtil: PageQueryUtil): List<Blog> {
101.        // 省略部分代码……
102.        return queryFactory.selectFrom(qBlog)
103.                .where(qBlog.blogStatus.eq(1).and(qBlog.isDeleted.eq(0)))
104.                .where(qBlog.blogId.`in`(blogIds))
105.                .orderBy(OrderSpecifier(Order.DESC, qBlog.blogId))
106.                .offset(start.toLong())
107.                .limit(limit.toLong())
108.                .fetchResults()
109.                .results
110.    }
111.    // 根据 tagId 查询总数
112.    fun getTotalBlogsByTagId(pageUtil: PageQueryUtil): Int {
113.        // 省略部分代码……
114.        return queryFactory.selectFrom(qBlog)
115.                .where(qBlog.blogStatus.eq(1).and(qBlog.isDeleted.eq(0)))
116.                .where(qBlog.blogId.`in`(blogIds))
117.                .fetchCount().toInt()
118.    }
119.    // 根据 subUrl 查找 Blog
120.    fun selectBySubUrl(subUrl: String): Blog? {
121.        val qBlog = QBlog.blog
122.        return queryFactory.selectFrom(qBlog)
123.                .where(qBlog.blogSubUrl.eq(subUrl).and(qBlog.isDeleted.eq(0)))
124.                .limit(1)
125.                .fetchOne()
126.    }
```

```kotlin
127.    // 批量更新 categoryName 和 categoryId
128.    @Transactional
129.    fun updateBlogCategorys(categoryName: String, categoryId: Int, ids:
    List<Long>): Int {
130.        val qBlog = QBlog.blog
131.        return queryFactory.update(qBlog)
132.            .set(qBlog.blogCategoryId, categoryId)
133.            .set(qBlog.blogCategoryName, categoryName)
134.            .where(qBlog.blogId.`in`(ids).and(qBlog.isDeleted.eq(0)))
135.            .execute().toInt()
136.    }
137. }
```

BlogLinkDAO.kt 基于 QueryDSL 实现了根据 linkId 删除链接（友链）；插入一条友链；根据 linkId 查询友链；更新不为 null 的友链属性；根据友链 id 更新相关属性；按照友链 id 降序分页查找友链；查询友链总数；批量删除友链等。代码如下：

```kotlin
1.  @Component
2.  class BlogLinkDAO {
3.      @Autowired
4.      lateinit var queryFactory: JPAQueryFactory
5.      @Autowired
6.      lateinit var blogLinkRepository: BlogLinkRepository
7.      // 根据 linkId 删除 BlogLink
8.      @Transactional
9.      fun deleteByPrimaryKey(linkId: Int): Int {
10.         val qBlogLink = QBlogLink.blogLink
11.         return queryFactory.update(qBlogLink)
12.             .set(qBlogLink.isDeleted, 1)
13.             .where(qBlogLink.linkId.eq(linkId).and(qBlogLink.isDeleted.eq(0)))
14.             .execute()
15.             .toInt()
16.     }
17.     // 保存一条 BlogLink
18.     @Transactional
19.     fun insert(record: BlogLink): Int {
20.         blogLinkRepository.save(record)
```

```kotlin
21.         return 0
22.     }
23.     // 保存一条 BlogLink
24.     @Transactional
25.     fun insertSelective(record: BlogLink): Int {
26.         blogLinkRepository.save(record)
27.         return 0
28.     }
29.     // 根据 linkId 查找 BlogLink
30.     fun selectByPrimaryKey(linkId: Int?): BlogLink {
31.         val qBloglink = QBlogLink.blogLink
32.         return queryFactory.selectFrom(qBloglink)
33.                 .where(qBloglink.linkId.eq(linkId).and(qBloglink.isDeleted.eq(0)))
34.                 .fetchFirst()
35.     }
36.     // 将 record 不为 null 的字段更新到数据库
37.     @Transactional
38.     fun updateByPrimaryKeySelective(record: BlogLink): Int {
39.         // 省略部分代码……
40.         return queryFactory.update(qBlogLink)
41.                 .set(cols, values)
42.                 .where(qBlogLink.linkId.eq(record.linkId))
43.                 .execute()
44.                 .toInt()
45.     }
46.     // 更新 BlogLink
47.     @Transactional
48.     fun updateByPrimaryKey(record: BlogLink): Int {
49.         // 省略部分代码……
50.     }
51.     // 分页查找 BlogLink
52.     fun findLinkList(pageUtil: PageQueryUtil?): List<BlogLink> {
53.         // 省略部分代码……
54.         return return queryFactory.selectFrom(qBlogLink)
55.                 .where(qBlogLink.isDeleted.eq(0))
56.                 .orderBy(OrderSpecifier(Order.DESC, qBlogLink.linkId))
57.                 .fetchResults()
```

```
58.             .results
59.     }
60.     // 查询总数
61.     fun getTotalLinks(pageUtil: PageQueryUtil?): Int {
62.         // 省略部分代码……
63.     }
64.     // 根据 ids 批量删除
65.     @Transactional
66.     fun deleteBatch(ids: List<Int>): Int {
67.         // 省略部分代码……
68.     }
69. }
```

BlogTagDAO.kt 基于 QueryDSL 实现了根据标签 id 删除标签、插入一条标签、根据 tagId 查询标签、根据 tagName 查询标签、选择性更新不为 null 的标签属性、根据 tagId 更新标签名称、创建时间、按照 tagId 降序分页查找友链、查询不同标签对应的博客的数量、查询标签总数、批量删除标签、批量保存标签等。代码如下：

```
1.  @Component
2.  class BlogTagDAO {
3.      @Autowired
4.      lateinit var queryFactory: JPAQueryFactory
5.      @Autowired
6.      lateinit var blogTagRepository: BlogTagRepository
7.      // 根据 tagId 删除 BlogTag
8.      @Transactional
9.      fun deleteByPrimaryKey(tagId: Int): Int {
10.         val qBlogTag = QBlogTag.blogTag
11.         return queryFactory.update(qBlogTag)
12.             .set(qBlogTag.isDeleted, 1)
13.             .where(qBlogTag.tagId.eq(tagId))
14.             .execute().toInt()
15.     }
16.     // 保存一条 BlogTag
17.     @Transactional
18.     fun insert(record: BlogTag): Int {
19.         blogTagRepository.save(record)
```

```kotlin
20.        return 0
21.    }
22.    // 保存一条 BlogTag
23.    @Transactional
24.    fun insertSelective(record: BlogTag): Int {
25.        blogTagRepository.save(record)
26.        return 0
27.    }
28.    // 根据 tagId 查询 BlogTag
29.    fun selectByPrimaryKey(tagId: Int): BlogTag {
30.        val qBlogTag = QBlogTag.blogTag
31.        return queryFactory.selectFrom(qBlogTag)
32.                .where(qBlogTag.tagId.eq(tagId).and(qBlogTag.isDeleted.eq(0)))
33.                .fetchFirst()
34.    }
35.    // 根据 tagName 查询 BlogTag
36.    fun selectByTagName(tagName: String): BlogTag {
37.        val qBlogTag = QBlogTag.blogTag
38.        return queryFactory.selectFrom(qBlogTag)
39.                .where(qBlogTag.tagName.eq(tagName).and(qBlogTag.isDeleted.eq(0)))
40.                .fetchFirst()
41.    }
42.    // 将 record 不为 null 的字段更新到数据库
43.    @Transactional
44.    fun updateByPrimaryKeySelective(record: BlogTag): Int {
45.        // 省略部分代码……
46.        return queryFactory.update(qBlogTag)
47.                .set(cols, values)
48.                .where(qBlogTag.tagId.eq(record.tagId))
49.                .execute()
50.                .toInt()
51.    }
52.    // 更新 BlogTag
53.    @Transactional
54.    fun updateByPrimaryKey(record: BlogTag): Int {
55.        // 省略部分代码……
56.    }
```

```kotlin
57.    // 分页查找 BlogTag
58.    fun findTagList(pageUtil: PageQueryUtil): List<BlogTag> {
59.        // 省略部分代码……
60.    }
61.    // 查询每个 tagId、tagName 对应的 Blog 数量
62.    fun getTagCount(): List<BlogTagCount> {
63.        // 省略部分代码……
64.        return result.map {
65.            val blogTagCount = BlogTagCount()
66.            blogTagCount.tagId = it.get(0, Int::class.java)!!
67.            blogTagCount.tagName = it.get(1, String::class.java)!!
68.            blogTagCount.tagCount = it.get(2, Long::class.java)?.toInt()!!
69.            blogTagCount
70.        }.toList()
71.    }
72.    // 查询 BlogTag 总数
73.    fun getTotalTags(pageUtil: PageQueryUtil?): Int {
74.        val qBlogTag = QBlogTag.blogTag
75.        return queryFactory.selectFrom(qBlogTag)
76.                .fetchCount().toInt()
77.    }
78.    // 根据 ids 批量删除 BlogTag
79.    @Transactional
80.    fun deleteBatch(ids: List<Int>): Int {
81.        val qBlogTag = QBlogTag.blogTag
82.        return queryFactory.update(qBlogTag)
83.                .set(qBlogTag.isDeleted, 1)
84.                .where(qBlogTag.tagId.`in`(ids))
85.                .execute().toInt()
86.    }
87.    // 批量保存 BlogTag
88.    @Transactional
89.    fun batchInsertBlogTag(tagList: List<BlogTag>): Int {
90.        blogTagRepository.saveAll(tagList)
91.        return 0
92.    }
93. }
```

BlogTagRelationDAO.kt 基于 QueryDSL 实现了根据 relationId 删除记录；插入记录；根据 relationId 查找记录；根据 blogId、tagId 查找关系记录；根据标签列表查询关系记录；选择性更新不为 null 的记录的属性；更新记录的属性；批量保存关系记录；批量删除关系记录等。代码如下：

```
1.  @Component
2.  class BlogTagRelationDAO {
3.      @Autowired
4.      lateinit var queryFactory: JPAQueryFactory
5.      @Autowired
6.      lateinit var blogTagRelationRepository: BlogTagRelationRepository
7.      // 根据 relationId 删除记录
8.      @Transactional
9.      fun deleteByPrimaryKey(relationId: Long): Int {
10.         val qBlogTagRelation = QBlogTagRelation.blogTagRelation
11.         return queryFactory.delete(qBlogTagRelation)
12.             .where(qBlogTagRelation.relationId.eq(relationId))
13.             .execute().toInt()
14.     }
15.     // 保存数据
16.     @Transactional
17.     fun insert(record: BlogTagRelation): Int {
18.         blogTagRelationRepository.save(record)
19.         return 0
20.     }
21.     // 保存数据
22.     @Transactional
23.     fun insertSelective(record: BlogTagRelation): Int {
24.         blogTagRelationRepository.save(record)
25.         return 0
26.     }
27.     // 根据 relationId 查找
28.     fun selectByPrimaryKey(relationId: Long): BlogTagRelation {
29.         val qBlogTagRelation = QBlogTagRelation.blogTagRelation
30.         return queryFactory.selectFrom(qBlogTagRelation)
31.             .where(qBlogTagRelation.relationId.eq(relationId))
```

```kotlin
32.            .fetchFirst()
33.    }
34.    // 根据 blogId、tagId 查找
35.    fun selectByBlogIdAndTagId(blogId: Long, tagId: Int): BlogTagRelation {
36.        val qBlogTagRelation = QBlogTagRelation.blogTagRelation
37.        return queryFactory.selectFrom(qBlogTagRelation)
38.                .where(qBlogTagRelation.blogId.eq(blogId).and(qBlogTagRelation.tagId.eq(tagId)))
39.                .fetchFirst()
40.    }
41.    // 查找表中存在的 tagIds
42.    fun selectDistinctTagIds(tagIds: List<Int>): List<Int> {
43.        val qBlogTagRelation = QBlogTagRelation.blogTagRelation
44.        return queryFactory.selectDistinct(qBlogTagRelation.tagId)
45.                .from(qBlogTagRelation)
46.                .where(qBlogTagRelation.tagId.`in`(tagIds))
47.                .fetchResults()
48.                .results
49.    }
50.    // 将 record 不为 null 的字段更新到数据库
51.    @Transactional
52.    fun updateByPrimaryKeySelective(record: BlogTagRelation): Int {
53.        // 省略部分代码……
54.        return queryFactory.update(qBlogTagRelation)
55.                .set(cols, values)
56.                .where(qBlogTagRelation.relationId.eq(record.relationId))
57.                .execute()
58.                .toInt()
59.    }
60.    // 更新 BlogTagRelation
61.    @Transactional
62.    fun updateByPrimaryKey(record: BlogTagRelation): Int {
63.        // 省略部分代码……
64.    }
65.    // 批量插入 BlogTagRelation
66.    @Transactional
67.    fun batchInsert(blogTagRelationList: List<BlogTagRelation>): Int {
```

```
68.         blogTagRelationRepository.saveAll(blogTagRelationList)
69.         return 0
70.     }
71.     // 根据 blogId 删除
72.     @Transactional
73.     fun deleteByBlogId(blogId: Long?): Int {
74.         val qBlogTagRelation = QBlogTagRelation.blogTagRelation
75.         return queryFactory.delete(qBlogTagRelation)
76.                 .where(qBlogTagRelation.blogId.eq(blogId))
77.                 .execute().toInt()
78.     }
79. }
```

10.6 Service 层的设计

博客定义的服务层如表 10.4 所示。

表 10.4 博客定义的服务列表

序号	服务	说明
1	AdminUserService	管理用户服务
2	BlogService	博客服务
3	CategoryService	分类服务
4	CommentService	评论服务
5	ConfigService	配置服务
6	LinkService	链接（友链）服务
7	TagService	标签服务

AdminUserService.kt 提供了管理用户相关服务：登录、获取用户信息、修改当前登录用户的密码及修改当前登录用户的名称信息。代码如下：

```
1. interface AdminUserService {
2.     // 登录
3.     fun login(userName: String, password: String): AdminUser?
4.     /**
5.      * 获取用户信息
6.      * @param loginUserId
```

```kotlin
7.      * @return
8.      */
9.     fun getUserDetailById(loginUserId: Int): AdminUser?
10.    /**
11.     * 修改当前登录用户的密码
12.     * @param loginUserId
13.     * @param originalPassword
14.     * @param newPassword
15.     * @return
16.     */
17.    fun updatePassword(loginUserId: Int, originalPassword: String, newPassword: String): Boolean
18.    /**
19.     * 修改当前登录用户的名称信息
20.     * @param loginUserId
21.     * @param loginUserName
22.     * @param nickName
23.     * @return
24.     */
25.    fun updateName(loginUserId: Int, loginUserName: String, nickName: String): Boolean
26. }
```

BlogService.kt 提供了博客相关服务：保存博客内容、分页查询博客、批量删除博客、获取博客总数、根据 id 获取详情、更新博客、获取首页文章列表、查询首页侧边栏显示的博客、获取文章详情、根据分类获取文章列表及根据搜索框的输入获取文章列表等。代码如下：

```kotlin
1. interface BlogService {
2.     // 保存博客
3.     fun saveBlog(blog: Blog): String
4.     // 分页查询博客
5.     fun getBlogsPage(pageUtil: PageQueryUtil): PageResult
6.     // 批量删除博客
7.     fun deleteBatch(ids: List<Long>): Boolean
8.     // 获取博客总数
9.     fun getTotalBlogs(): Int
```

```kotlin
10.    /**
11.     * 根据id获取详情
12.     * @param blogId
13.     * @return
14.     */
15.    fun getBlogById(blogId: Long): Blog
16.    /**
17.     * 后台修改
18.     * @param blog
19.     * @return
20.     */
21.    fun updateBlog(blog: Blog): String
22.    /**
23.     * 获取首页文章列表
24.     * @param page
25.     * @return
26.     */
27.    fun getBlogsForIndexPage(page: Int): PageResult
28.    /**
29.     * 获取首页侧边栏数据列表
30.     * 0-点击最多 1-最新发布
31.     * @param type
32.     * @return
33.     */
34.    fun getBlogListForIndexPage(type: Int): List<SimpleBlogListVO>
35.    /**
36.     * 文章详情
37.     * @param blogId
38.     * @return
39.     */
40.    fun getBlogDetail(blogId: Long): BlogDetailVO?
41.    /**
42.     * 根据标签获取文章列表
43.     * @param tagName
44.     * @param page
45.     * @return
46.     */
```

```kotlin
47.     fun getBlogsPageByTag(tagName: String, page: Int): PageResult?
48.     /**
49.      * 根据分类获取文章列表
50.      * @param categoryId
51.      * @param page
52.      * @return
53.      */
54.     fun getBlogsPageByCategory(categoryId: String, page: Int): PageResult?
55.     /**
56.      * 根据搜索获取文章列表
57.      * @param keyword
58.      * @param page
59.      * @return
60.      */
61.     fun getBlogsPageBySearch(keyword: String, page: Int): PageResult?
62.     // 根据 subUrl 查找博客明细
63.     fun getBlogDetailBySubUrl(subUrl: String): BlogDetailVO?
64. }
```

CategoryService.kt 提供了分类相关服务：分页查询分类信息及添加分类数据等。代码如下：

```kotlin
1.  interface CategoryService {
2.      /**
3.       * 查询分类的分页数据
4.       * @param pageUtil
5.       * @return
6.       */
7.      fun getBlogCategoryPage(pageUtil: PageQueryUtil): PageResult
8.      // 查询分类总数
9.      fun getTotalCategories(): Int
10.     /**
11.      * 添加分类数据
12.      * @param categoryName
13.      * @param categoryIcon
14.      * @return
15.      */
```

```
16.     fun saveCategory(categoryName: String, categoryIcon: String): Boolean
17.     // 更新分类名称及图标
18.     fun updateCategory(categoryId: Int, categoryName: String, categoryIcon:
String): Boolean
19.     // 批量删除
20.     fun deleteBatch(ids: List<Int>): Boolean
21.     // 查找所有分类
22.     fun getAllCategories(): List<BlogCategory>
23. }
```

CommentService.kt 提供了评论相关服务：添加评论、分页查询评论、批量审核评论、批量删除评论、对评论添加回复、根据文章 id 和分页参数获取文章的评论列表等。代码如下：

```
1.  interface CommentService {
2.      /**
3.       * 添加评论
4.       * @param blogComment
5.       * @return
6.       */
7.      fun addComment(blogComment: BlogComment): Boolean
8.      /**
9.       * 后台管理系统中的评论分页功能
10.      * @param pageUtil
11.      * @return
12.      */
13.     fun getCommentsPage(pageUtil: PageQueryUtil): PageResult
14.     // 查询评论总数
15.     fun getTotalComments(): Int
16.     /**
17.      * 批量审核
18.      * @param ids
19.      * @return
20.      */
21.     fun checkDone(ids: List<Long>): Boolean
22.     /**
23.      * 批量删除
24.      * @param ids
```

```kotlin
25.     * @return
26.     */
27.    fun deleteBatch(ids: List<Long>): Boolean
28.    /**
29.     * 添加回复
30.     * @param commentId
31.     * @param replyBody
32.     * @return
33.     */
34.    fun reply(commentId: Long, replyBody: String): Boolean
35.    /**
36.     * 根据文章 id 和分页参数获取文章的评论列表
37.     * @param blogId
38.     * @param page
39.     * @return
40.     */
41.    fun getCommentPageByBlogIdAndPageNum(blogId: Long, page: Int): PageResult?
42. }
```

ConfigService.kt 提供了配置相关服务：修改配置项及获取所有的配置项等。代码如下：

```kotlin
1. interface ConfigService {
2.    /**
3.     * 修改配置项
4.     * @param configName
5.     * @param configValue
6.     * @return
7.     */
8.    fun updateConfig(configName: String, configValue: String): Int
9.    /**
10.     * 获取所有的配置项
11.     * @return
12.     */
13.    fun getAllConfigs(): Map<String, String>
14. }
```

LinkService.kt 提供了友链相关服务：分页查询友链、查询友链总数、保存友链、根据 id 查询友链、更新友链内容、批量删除友链、返回友链页面所需的所有数据等。代码如下：

```kotlin
1.  interface LinkService {
2.      /**
3.       * 查询友链的分页数据
4.       * @param pageUtil
5.       * @return
6.       */
7.      fun getBlogLinkPage(pageUtil: PageQueryUtil): PageResult
8.      // 获取友链总数
9.      fun getTotalLinks(): Int
10.     // 保存友链
11.     fun saveLink(link: BlogLink): Boolean
12.     // 根据 id 查询友链
13.     fun selectById(id: Int?): BlogLink
14.     // 更新友链
15.     fun updateLink(tempLink: BlogLink): Boolean
16.     // 批量删除
17.     fun deleteBatch(ids: List<Int>): Boolean
18.     /**
19.      * 返回友链页面所需的所有数据
20.      * @return
21.      */
22.     fun getLinksForLinkPage(): Map<Byte, List<BlogLink>>
23. }
```

TagService.kt 提供了标签服务：分页查询标签数据、获取标签总数、保存标签、批量删除标签及获取每个标签对应的博客总数等。代码如下：

```kotlin
1.  interface TagService {
2.      /**
3.       * 查询标签的分页数据
4.       * @param pageUtil
5.       * @return
6.       */
7.      fun getBlogTagPage(pageUtil: PageQueryUtil): PageResult
8.      // 查询标签总数
9.      fun getTotalTags(): Int
10.     // 保存标签
```

```
11.     fun saveTag(tagName: String): Boolean
12.     // 批量删除
13.     fun deleteBatch(ids: List<Int>): Boolean
14.     // 获取每个标签对应的博客总数
15.     fun getBlogTagCountForIndex(): List<BlogTagCount>
16. }
```

10.7 Controller 层的设计

博客定义的接口层如表 10.5 所示。

表 10.5 博客定义的接口列表

序号	controller	说明
前台接口		
1	MyBlogController	提供博客前台展示相关接口
后台接口		
2	AdminController	后台管理相关接口
3	BlogController	博客管理相关接口
4	CategoryController	类别管理相关接口
5	CommentController	评论管理相关接口
6	ConfigurationController	配置管理相关接口
7	LinkController	链接管理相关接口
8	TagController	标签管理相关接口
9	UploadController	上传相关接口

AdminController.kt 提供了后台管理相关接口和视图跳转：登录、跳转到管理后台首页、获取用户信息、修改密码、修改用户名称及登出等。代码如下：

```
1.  @Controller
2.  @RequestMapping("/admin")
3.  class AdminController {
4.      @Resource
5.      lateinit var adminUserService: AdminUserService
6.      @Resource
7.      lateinit var blogService: BlogService
8.      @Resource
```

```kotlin
9.      lateinit var categoryService: CategoryService
10.     @Resource
11.     lateinit var linkService: LinkService
12.     @Resource
13.     lateinit var tagService: TagService
14.     @Resource
15.     lateinit var commentService: CommentService
16.     // 跳转到登录页
17.     @GetMapping("/login")
18.     fun login(): String {
19.         return "admin/login"
20.     }
21.     // 跳转到管理后台首页
22.     @GetMapping("", "/", "/index", "/index.html")
23.     fun index(request: HttpServletRequest): String {
24.         // 省略部分代码……
25.         return "admin/index"
26.     }
27.     // 登录
28.     @PostMapping(value = ["/login"])
29.     fun login(@RequestParam("userName") userName: String,
30.               @RequestParam("password") password: String,
31.               @RequestParam("verifyCode") verifyCode: String,
32.               session: HttpSession): String {
33.         if (StringUtils.isEmpty(verifyCode)) {
34.             session.setAttribute("errorMsg", "验证码不能为空")
35.             return "admin/login"
36.         }
37.         if (StringUtils.isEmpty(userName) || StringUtils.isEmpty(password)) {
38.             session.setAttribute("errorMsg", "用户名或密码不能为空")
39.             return "admin/login"
40.         }
41.         val kaptchaCode = session?.getAttribute("verifyCode").toString() + ""
42.         if (StringUtils.isEmpty(kaptchaCode) || verifyCode != kaptchaCode) {
43.             session.setAttribute("errorMsg", "验证码错误")
44.             return "admin/login"
```

```kotlin
45.        }
46.        val adminUser = adminUserService.login(userName, password)
47.        if (adminUser != null) {
48.            session.setAttribute("loginUser", adminUser.nickName)
49.            session.setAttribute("loginUserId", adminUser.adminUserId)
50.            //将session过期时间设置为7200秒,即两小时
51.            //session.setMaxInactiveInterval(60 * 60 * 2);
52.            return "redirect:/admin/index"
53.        } else {
54.            session.setAttribute("errorMsg", "登录失败")
55.            return "admin/login"
56.        }
57.    }
58.    // 跳转到个人信息视图页
59.    @GetMapping("/profile")
60.    fun profile(request: HttpServletRequest): String {
61.        // 省略部分代码……
62.        return "admin/profile"
63.    }
64.    // 更新用户密码
65.    @PostMapping("/profile/password")
66.    @ResponseBody
67.    fun passwordUpdate(request: HttpServletRequest,
    @RequestParam("originalPassword") originalPassword: String,
68.                       @RequestParam("newPassword") newPassword: String): String {
69.        // 省略部分代码……
70.    }
71.    // 更新用户名
72.    @PostMapping("/profile/name")
73.    @ResponseBody
74.    fun nameUpdate(request: HttpServletRequest, @RequestParam("loginUserName")
    loginUserName: String,
75.                   @RequestParam("nickName") nickName: String): String {
76.        // 省略部分代码……
77.    }
78.    // 跳转到登出视图页
```

```kotlin
79.     @GetMapping("/logout")
80.     fun logout(request: HttpServletRequest): String {
81.         // 省略部分代码……
82.         return "admin/login"
83.     }
84. }
```

BlogController.kt 提供了博客管理相关接口和视图跳转：展示博客列表、跳转到博客管理页面、编辑博客、保存博客、更新博客、上传博客封面图片及删除博客等。代码如下：

```kotlin
1.  @Controller
2.  @RequestMapping("/admin")
3.  class BlogController {
4.      @Resource
5.      lateinit var blogService: BlogService
6.      @Resource
7.      lateinit var categoryService: CategoryService
8.      // 获取博客列表
9.      @GetMapping("/blogs/list")
10.     @ResponseBody
11.     fun list(@RequestParam params: Map<String, Any>): Result<Any> {
12.         // 省略部分代码……
13.     }
14.     // 跳转到博客视图页
15.     @GetMapping("/blogs")
16.     fun list(request: HttpServletRequest): String {
17.         request.setAttribute("path", "blogs")
18.         return "admin/blog"
19.     }
20.     // 跳转到博客编辑视图页
21.     @GetMapping("/blogs/edit")
22.     fun edit(request: HttpServletRequest): String {
23.         // 省略部分代码……
24.         return "admin/edit"
25.     }
26.     // 跳转到博客编辑视图页
27.     @GetMapping("/blogs/edit/{blogId}")
```

```kotlin
28.    fun edit(request: HttpServletRequest, @PathVariable("blogId") blogId: Long):
   String {
29.        // 省略部分代码……
30.        return "admin/edit"
31.    }
32.    // 保存博客
33.    @PostMapping("/blogs/save")
34.    @ResponseBody
35.    fun save(@RequestParam("blogTitle") blogTitle: String,
36.             @RequestParam(name = "blogSubUrl", required = false) blogSubUrl: String,
37.             @RequestParam("blogCategoryId") blogCategoryId: Int,
38.             @RequestParam("blogTags") blogTags: String,
39.             @RequestParam("blogContent") blogContent: String,
40.             @RequestParam("blogCoverImage") blogCoverImage: String,
41.             @RequestParam("blogStatus") blogStatus: Byte,
42.             @RequestParam("enableComment") enableComment: Byte): Result<Any> {
43.        // 省略部分代码……
44.        val saveBlogResult = blogService.saveBlog(blog)
45.        return if ("success" == saveBlogResult) {
46.            ResultGenerator.genSuccessResult("添加成功")
47.        } else {
48.            ResultGenerator.genFailResult(saveBlogResult)
49.        }
50.    }
51.    // 更新博客
52.    @PostMapping("/blogs/update")
53.    @ResponseBody
54.    fun update(@RequestParam("blogId") blogId: Long,
55.               @RequestParam("blogTitle") blogTitle: String,
56.               @RequestParam(name = "blogSubUrl", required = false) blogSubUrl: String,
57.               @RequestParam("blogCategoryId") blogCategoryId: Int,
58.               @RequestParam("blogTags") blogTags: String,
59.               @RequestParam("blogContent") blogContent: String,
60.               @RequestParam("blogCoverImage") blogCoverImage: String,
61.               @RequestParam("blogStatus") blogStatus: Byte,
62.               @RequestParam("enableComment") enableComment: Byte): Result<Any> {
63.        // 省略部分代码……
```

```
64.        val updateBlogResult = blogService.updateBlog(blog)
65.        return if ("success" == updateBlogResult) {
66.            ResultGenerator.genSuccessResult("修改成功")
67.        } else {
68.            ResultGenerator.genFailResult(updateBlogResult)
69.        }
70.    }
71.    // 上传图片
72.    @PostMapping("/blogs/md/uploadfile")
73.    @Throws(IOException::class, URISyntaxException::class)
74.    fun uploadFileByEditormd(request: HttpServletRequest,
75.                             response: HttpServletResponse,
76.                             @RequestParam(name = "editormd-image-file", required = true)
77.                             file: MultipartFile) {
78.        // 省略部分代码……
79.    }
80.    // 批量删除博客
81.    @PostMapping("/blogs/delete")
82.    @ResponseBody
83.    fun delete(@RequestBody ids: List<Long>): Result<Any> {
84.        if (ids.size < 1) {
85.            return ResultGenerator.genFailResult("参数异常！")
86.        }
87.        return if (blogService.deleteBatch(ids)) {
88.            ResultGenerator.genSuccessResult()
89.        } else {
90.            ResultGenerator.genFailResult("删除失败")
91.        }
92.    }
93. }
```

CategoryController.kt 提供类别管理接口和视图跳转：获取分类列表、添加分类、修改分类、删除分类、跳转到分类管理页面等。代码如下：

```
1. @Controller
2. @RequestMapping("/admin")
3. class CategoryController {
```

```kotlin
4.      @Resource
5.      lateinit var categoryService: CategoryService
6.      // 跳转到类别视图页
7.      @GetMapping("/categories")
8.      fun categoryPage(request: HttpServletRequest): String {
9.          request.setAttribute("path", "categories")
10.         return "admin/category"
11.     }
12.     /**
13.      * 分类列表
14.      */
15.     @RequestMapping(value = ["/categories/list"], method = [RequestMethod.GET])
16.     @ResponseBody
17.     fun list(@RequestParam params: Map<String, Any>): Result<Any> {
18.         if (StringUtils.isEmpty(params["page"]) || StringUtils.isEmpty(params["limit"])) {
19.             return ResultGenerator.genFailResult("参数异常！")
20.         }
21.         val pageUtil = PageQueryUtil(params)
22.         return ResultGenerator.genSuccessResult(categoryService!!.getBlogCategoryPage(pageUtil))
23.     }
24.     /**
25.      * 添加分类
26.      */
27.     @RequestMapping(value = ["/categories/save"], method = [RequestMethod.POST])
28.     @ResponseBody
29.     fun save(@RequestParam("categoryName") categoryName: String,
30.              @RequestParam("categoryIcon") categoryIcon: String): Result<Any> {
31.         // 省略部分代码……
32.     }
33.     /**
34.      * 修改分类
35.      */
36.     @RequestMapping(value = ["/categories/update"], method = [RequestMethod.POST])
37.     @ResponseBody
38.     fun update(@RequestParam("categoryId") categoryId: Int,
```

```kotlin
39.                @RequestParam("categoryName") categoryName: String,
40.                @RequestParam("categoryIcon") categoryIcon: String): Result<Any> {
41.        // 省略部分代码……
42.    }
43.    /**
44.     * 删除分类
45.     */
46.    @RequestMapping(value = ["/categories/delete"], method = [RequestMethod.POST])
47.    @ResponseBody
48.    fun delete(@RequestBody ids: List<Int>): Result<Any> {
49.        // 省略部分代码……
50.    }
51. }
```

CommentController.kt 提供了评论管理接口和视图跳转：获取评论列表、审核评论、回复评论、删除评论、跳转到评论管理页面等。代码如下：

```kotlin
1.  @Controller
2.  @RequestMapping("/admin")
3.  class CommentController {
4.      @Resource
5.      lateinit var commentService: CommentService
6.      // 获取评论列表
7.      @GetMapping("/comments/list")
8.      @ResponseBody
9.      fun list(@RequestParam params: Map<String, Any>): Result<Any> {
10.         // 省略部分代码……
11.     }
12.     // 审核评论
13.     @PostMapping("/comments/checkDone")
14.     @ResponseBody
15.     fun checkDone(@RequestBody ids: List<Long>): Result<Any> {
16.         // 省略部分代码……
17.     }
18.     // 回复评论
19.     @PostMapping("/comments/reply")
20.     @ResponseBody
```

```kotlin
21.    fun checkDone(@RequestParam("commentId") commentId: Long?,
22.                  @RequestParam("replyBody") replyBody: String): Result<Any> {
23.        // 省略部分代码……
24.    }
25.    // 删除评论
26.    @PostMapping("/comments/delete")
27.    @ResponseBody
28.    fun delete(@RequestBody ids: List<Long>): Result<Any> {
29.        // 省略部分代码……
30.    }
31.    // 跳转到评论视图页
32.    @GetMapping("/comments")
33.    fun list(request: HttpServletRequest): String {
34.        request.setAttribute("path", "comments")
35.        return "admin/comment"
36.    }
37. }
```

ConfigurationController.kt 提供了配置管理接口和视图跳转：跳转到配置管理页面、更新网站相关配置、更新用户相关配置及更新网站页脚相关配置等。代码如下：

```kotlin
1.  @Controller
2.  @RequestMapping("/admin")
3.  class ConfigurationController {
4.      @Resource
5.      lateinit var configService: ConfigService
6.      // 跳转到配置视图页
7.      @GetMapping("/configurations")
8.      fun list(request: HttpServletRequest): String {
9.          // 省略部分代码……
10.         return "admin/configuration"
11.     }
12.     // 获取博客 website 配置
13.     @PostMapping("/configurations/website")
14.     @ResponseBody
15.     fun website(@RequestParam(value = "websiteName", required = false) websiteName: String,
```

```
16.             @RequestParam(value = "websiteDescription", required = false)
    websiteDescription: String,
17.             @RequestParam(value = "websiteLogo", required = false) websiteLogo:
    String,
18.             @RequestParam(value = "websiteIcon", required = false) websiteIcon:
    String): Result<Any> {
19.        // 省略部分代码……
20.    }
21.    //获取博客 userInfo 配置
22.    @PostMapping("/configurations/userInfo")
23.    @ResponseBody
24.    fun userInfo(@RequestParam(value = "yourAvatar", required = false) yourAvatar:
    String,
25.             @RequestParam(value = "yourName", required = false) yourName:
    String,
26.             @RequestParam(value = "yourEmail", required = false) yourEmail:
    String): Result<Any> {
27.        // 省略部分代码……
28.    }
29.    // 获取博客 footer 配置
30.    @PostMapping("/configurations/footer")
31.    @ResponseBody
32.    fun footer(@RequestParam(value = "footerAbout", required = false) footerAbout:
    String,
33.             @RequestParam(value = "footerICP", required = false) footerICP:
    String,
34.             @RequestParam(value = "footerCopyRight", required = false)
    footerCopyRight: String,
35.             @RequestParam(value = "footerPoweredBy", required = false)
    footerPoweredBy: String,
36.             @RequestParam(value = "footerPoweredByURL", required = false)
    footerPoweredByURL: String): Result<Any> {
37.        // 省略部分代码…
38.    }
39. }
```

LinkController.kt 提供链接（友链）管理相关接口和视图跳转：跳转到友链管理页面、

获取友链列表、添加友链、获取友链详情、修改友链及删除友链等。代码如下：

```kotlin
1.  @Controller
2.  @RequestMapping("/admin")
3.  class LinkController {
4.      @Resource
5.      lateinit var linkService: LinkService
6.      // 跳转到友链视图
7.      @GetMapping("/links")
8.      fun linkPage(request: HttpServletRequest): String {
9.          request.setAttribute("path", "links")
10.         return "admin/link"
11.     }
12.     // 获取友链列表
13.     @GetMapping("/links/list")
14.     @ResponseBody
15.     fun list(@RequestParam params: Map<String, Any>): Result<Any> {
16.         // 省略部分代码……
17.     }
18.     /**
19.      * 添加友链
20.      */
21.     @RequestMapping(value = ["/links/save"], method = [RequestMethod.POST])
22.     @ResponseBody
23.     fun save(@RequestParam("linkType") linkType: Int?,
24.              @RequestParam("linkName") linkName: String,
25.              @RequestParam("linkUrl") linkUrl: String,
26.              @RequestParam("linkRank") linkRank: Int?,
27.              @RequestParam("linkDescription") linkDescription: String): Result<Any> {
28.         // 省略部分代码……
29.     }
30.     /**
31.      * 详情
32.      */
33.     @GetMapping("/links/info/{id}")
34.     @ResponseBody
35.     fun info(@PathVariable("id") id: Int?): Result<Any> {
```

```kotlin
36.         // 省略部分代码……
37.     }
38.     /**
39.      * 修改友链
40.      */
41.     @RequestMapping(value = ["/links/update"], method = [RequestMethod.POST])
42.     @ResponseBody
43.     fun update(@RequestParam("linkId") linkId: Int?,
44.                @RequestParam("linkType") linkType: Int?,
45.                @RequestParam("linkName") linkName: String,
46.                @RequestParam("linkUrl") linkUrl: String,
47.                @RequestParam("linkRank") linkRank: Int?,
48.                @RequestParam("linkDescription") linkDescription: String):
    Result<Any> {
49.         // 省略部分代码……
50.     }
51.     /**
52.      * 删除友链
53.      */
54.     @RequestMapping(value = ["/links/delete"], method = [RequestMethod.POST])
55.     @ResponseBody
56.     fun delete(@RequestBody ids: List<Int>): Result<Any> {
57.         // 省略部分代码……
58.     }
59. }
```

TagController.kt 提供了标签管理接口和视图跳转：跳转到标签管理页面、获取标签列表、保存标签及删除标签等。代码如下：

```kotlin
1. @Controller
2. @RequestMapping("/admin")
3. class TagController {
4.     @Resource
5.     lateinit var tagService: TagService
6.     // 跳转到标签视图
7.     @GetMapping("/tags")
8.     fun tagPage(request: HttpServletRequest): String {
```

```kotlin
9.          request.setAttribute("path", "tags")
10.         return "admin/tag"
11.     }
12.     // 获取标签列表
13.     @GetMapping("/tags/list")
14.     @ResponseBody
15.     fun list(@RequestParam params: Map<String, Any>): Result<Any> {
16.         // 省略部分代码……
17.     }
18.     // 保存标签
19.     @PostMapping("/tags/save")
20.     @ResponseBody
21.     fun save(@RequestParam("tagName") tagName: String): Result<Any> {
22.         // 省略部分代码……
23.     }
24.     // 删除标签
25.     @PostMapping("/tags/delete")
26.     @ResponseBody
27.     fun delete(@RequestBody ids: List<Int>): Result<Any> {
28.         // 省略部分代码……
29.     }
30. }
```

UploadController.kt 提供了上传文件的接口：

```kotlin
1. @Controller
2. @RequestMapping("/admin")
3. class UploadController {
4.     // 上传文件
5.     @PostMapping("/upload/file")
6.     @ResponseBody
7.     @Throws(URISyntaxException::class)
8.     fun upload(httpServletRequest: HttpServletRequest, @RequestParam("file") file: MultipartFile): Result<Any> {
9.         // 省略部分代码……
10.     }
11. }
```

MyBlogController.kt 提供了前台展示相关接口和视图跳转：跳转到博客首页、跳转到分类页面、跳转到博客详情页面、跳转到标签列表页面、跳转到搜索列表页、跳转到友链页、添加文章评论等。代码如下：

```kotlin
@Controller
class MyBlogController {
    @Resource
    lateinit var blogService: BlogService
    @Resource
    lateinit var tagService: TagService
    @Resource
    lateinit var linkService: LinkService
    @Resource
    lateinit var commentService: CommentService
    @Resource
    lateinit var configService: ConfigService
    @Resource
    lateinit var categoryService: CategoryService
    /**
     * 首页
     * @return
     */
    @GetMapping("/", "/index", "index.html")
    fun index(request: HttpServletRequest): String {
        return this.page(request, 1)
    }
    /**
     * 首页 分页数据
     * @return
     */
    @GetMapping("/page/{pageNum}")
    fun page(request: HttpServletRequest, @PathVariable("pageNum") pageNum: Int): String {
        // 省略部分代码……
        return "blog/$theme/index"
    }
```

```kotlin
32.    /**
33.     * Categories 页面(包括分类数据和标签数据)
34.     * @return
35.     */
36.    @GetMapping("/categories")
37.    fun categories(request: HttpServletRequest): String {
38.        // 省略部分代码……
39.        return "blog/$theme/category"
40.    }
41.    /**
42.     * 详情页
43.     * @return
44.     */
45.    @GetMapping("/blog/{blogId}", "/article/{blogId}")
46.    fun detail(request: HttpServletRequest, @PathVariable("blogId") blogId: Long,
    @RequestParam(value = "commentPage", required = false, defaultValue = "1")
    commentPage: Int): String {
47.        // 省略部分代码……
48.        return "blog/$theme/detail"
49.    }
50.    /**
51.     * 标签列表页
52.     * @return
53.     */
54.    @GetMapping("/tag/{tagName}")
55.    fun tag(request: HttpServletRequest, @PathVariable("tagName") tagName: String):
    String {
56.        return tag(request, tagName, 1)
57.    }
58.    /**
59.     * 标签列表页
60.     * @return
61.     */
62.    @GetMapping("/tag/{tagName}/{page}")
63.    fun tag(request: HttpServletRequest, @PathVariable("tagName") tagName: String,
    @PathVariable("page") page: Int): String {
```

```
64.        // 省略部分代码……
65.        return "blog/$theme/list"
66.    }
67.    /**
68.     * 分类列表页
69.     * @return
70.     */
71.    @GetMapping("/category/{categoryName}")
72.    fun category(request: HttpServletRequest, @PathVariable("categoryName") categoryName: String): String {
73.        return category(request, categoryName, 1)
74.    }
75.    /**
76.     * 分类列表页
77.     * @return
78.     */
79.    @GetMapping("/category/{categoryName}/{page}")
80.    fun category(request: HttpServletRequest, @PathVariable("categoryName") categoryName: String, @PathVariable("page") page: Int): String {
81.        // 省略部分代码……
82.        return "blog/$theme/list"
83.    }
84.    /**
85.     * 搜索列表页
86.     * @return
87.     */
88.    @GetMapping("/search/{keyword}")
89.    fun search(request: HttpServletRequest, @PathVariable("keyword") keyword: String): String {
90.        return search(request, keyword, 1)
91.    }
92.    /**
93.     * 搜索列表页
94.     * @return
95.     */
96.    @GetMapping("/search/{keyword}/{page}")
```

```kotlin
97.     fun search(request: HttpServletRequest, @PathVariable("keyword") keyword:
    String, @PathVariable("page") page: Int): String {
98.         // 省略部分代码……
99.         return "blog/$theme/list"
100.    }
101.    /**
102.     * 友情链接页
103.     * @return
104.     */
105.    @GetMapping("/link")
106.    fun link(request: HttpServletRequest): String {
107.        // 省略部分代码……
108.        return "blog/$theme/link"
109.    }
110.    /**
111.     * 评论操作
112.     */
113.    @PostMapping(value = ["/blog/comment"])
114.    @ResponseBody
115.    fun comment(request: HttpServletRequest, session: HttpSession,
116.                @RequestParam blogId: Long?, @RequestParam verifyCode: String,
117.                @RequestParam commentator: String, @RequestParam email: String,
118.                @RequestParam websiteUrl: String, @RequestParam commentBody:
    String): Result<Any> {
119.        // 省略部分代码……
120.        return
    ResultGenerator.genSuccessResult(commentService.addComment(comment))
121.    }
122.    /**
123.     * 关于页面以及其他配置了 subUrl 的文章页
124.     * @return
125.     */
126.    @GetMapping("/{subUrl}")
127.    fun detail(request: HttpServletRequest, @PathVariable("subUrl") subUrl:
    String): String {
128.        // 省略部分代码……
```

```
129.     }
130.     companion object {
131.         //public static String theme = "default";
132.         //public static String theme = "yummy-jekyll";
133.         var theme = "amaze"
134.     }
135. }
```

10.8　部署到腾讯云

在腾讯云上安装 JDK 1.8 和 MySQL。将子项目 chapter11-blog 用 Maven 打包成一个 Jar 文件，通过 xFTP 工具上传到腾讯云主机，并上传配置文件，修改数据库连接配置，应用目录如图 10.3 所示。在 MySQL 数据库中建表，并初始化数据。运行如下命令，启动博客服务，如图 10.4 所示。

```
nohup java -jar chapter11-blog-0.0.1-SNAPSHOT.jar > blog.log &
```

图10.3　腾讯云应用目录

图10.4　博客启动日志

第 10 章 基于 Kotlin 和 Spring Boot 搭建博客

博客首页、友链页、详情页及后台管理页如图 10.5、图 10.6、图 10.7 及图 10.8 所示。

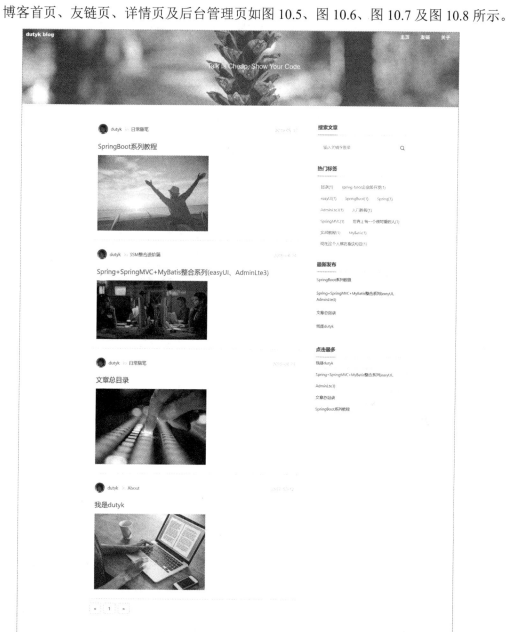

图10.5　博客首页

图10.6 博客友链页

图10.7 博客详情页

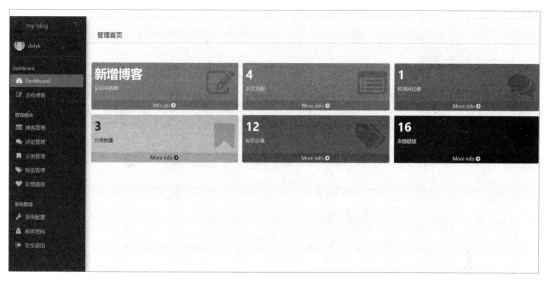

图10.8　博客管理后台

10.9　小结

本章通过一个博客例子介绍了如何使用 Kotlin、Spring Boot 开发一个简单的应用。本应用采用分层结构，自下而上分为数据实体层、数据库操作层、应用服务层及接口层。本章通过大量源码详细介绍了各层的开发方法，最后介绍了如何将博客应用部署到腾讯云，并展示了博客的效果。